THE MOLECULES OF LIFE

THE MOLECULES OF LIFE

DNA, RNA, and Proteins

RUSS HODGE

FOREWORD BY NADIA ROSENTHAL, PH.D.

This book is dedicated to the memory of my grandparents E. J. and Mabel Evens, to my parents, Ed and Jo Hodge, and especially to my wife, Gabi, and my children—Jesper, Sharon, and Lisa—with love.

THE MOLECULES OF LIFE: DNA, RNA, and Proteins

Facts On File, Inc.
An imprint of Infobase Publishing
132 West 31st Street
New York NY 10001

Library of Congress Cataloging-in-Publication Data
Hodge, Russ, 1961–
The molecules of life: DNA, RNA, and proteins / Russ Hodge ; foreword by Nadia Rosenthal.
p. cm.
Includes bibliographical references and index.
ISBN-13: 978-0-8160-6680-3
ISBN-10: 0-8160-6680-9
1. Nucleic acids—Popular works. 2. Proteins—Popular works. 3. Biochemistry—Popular works. 4. Molecular biology—Popular works. I. Title.
QP620.H63 2009
611'.01816—dc22 2008037094

Facts On File books are available at special discounts when purchased in bulk quantities for businesses, associations, institutions, or sales promotions. Please call our Special Sales Department in New York at (212) 967-8800 or (800) 322-8755.

You can find Facts On File on the World Wide Web at http://www.factsonfile.com

Text design by Kerry Casey
Illustrations by Richard Garratt
Photo research by Elizabeth H. Oakes

Printed in the United States of America

Bang Hermitage 10 9 8 7 6 5 4 3 2 1

This book is printed on acid-free paper.

I say that it touches a man that his blood is sea water and his tears are salt, that the seed of his loins is scarcely different from the same cells in a seaweed, and that of stuff like his bones coral is made. I say that the physical and biologic law lies down with him, and wakes when a child stirs in the womb, and that the sap in a tree, uprushing in the spring, and the smell of the loam, where the bacteria bestir themselves in darkness, and the path of the sun in the heaven, these are facts of first importance to his mental conclusions, and that a man who goes in no consciousness of them is a drifter and a dreamer, without a home or any contact with reality.

—Donald Culross Peattie,
An Almanac for Moderns: A Daybook of Nature, 1935 (1963)

Contents

Foreword

As a teaching assistant at Harvard University in the 1970s, I often sat in on George Wald's opening lecture for the undergraduate biology course. Wald was a towering figure in his field: His work on the molecule rhodopsin and its response to light in the retina had won him the Nobel Prize a few years earlier. Wald started a lecture by describing biology at the large end of the scale—man himself—then he methodically dissected the human organism into smaller and smaller constituents, picking apart the cell as the essential building block of the body and finally arriving at the molecules that make up our genes. "Now my friends," he concluded, "my survey may leave you with the disheartening impression that man is just a collection of molecules. That is perfectly true. However, my description is not meant to denigrate man, but rather to exalt the molecule!"

The Molecules of Life is just such an exaltation of molecules—those myriad components that come in all shapes, sizes, and functions, the parts and products of that giant machine, the cell. Considering that you need a microscope to see most cells, the study of the even tinier molecules that compose them has been largely indirect, based on observations of their properties or functions. As Russ Hodge explains in chapter 1, scientists during the last two centuries developed ingenious ways to probe the function of specific molecules, often without even a pure preparation to start from. When in the last 50 years the technique of X-ray crystallography was developed and scientists had their first direct glimpses of molecular structure, the pictures were so heavily coded that even today only an expert can decipher them.

In chapter 2 Hodge leads us through one of the most exciting scientific escapades in recent history: the discovery and description of the universal genetic material. DNA (deoxyribonucleic acid) is probably the most famous molecule in the life sciences and acts as both the central director of cell function and the repository of hereditary information in the tree of life. By bringing the field of genetics down from a rarified theoretical plane of predictable, mathematical precision to the level of messy, wet biochemistry, James Watson, Francis Crick, and their colleagues started a revolution in molecular biology that is still playing out in research institutes, pharmaceutical companies, and even forensic laboratories around the world. Molecules have gone mainstream.

How do cells use molecules? In chapter 3 the cell comes to life as it goes about its daily existence, either as a single bacterium, amoeba, egg, or sperm or as a member of a cellular commune in the tissues of a much larger organism. We learn about cells' ingenious manufacture and deployment of special molecules to probe and respond to their surroundings, to signal to their neighbors, or to relay information encoded in electric or chemical currents.

A constant flurry of molecular activity maintains our bodies even when at rest, due to the tireless work of the cells that form our bodies. Continuously managing their internal health and metabolism requires cells to change shape and sometimes to move actively about the body. Chapter 4 captures this frenzied scene and introduces specific molecules responsible for coordinating and orchestrating the action.

The final chapter examines current hot topics in clinical studies of biomolecules and gives a view into possible future medical applications of our knowledge as it grows. A molecular understanding of a disease is critical to the design of effective cures. Hodge explains how our immune systems are in a constant battle to distinguish those molecules that belong in our bodies—namely, our own—from foreign pathogenic molecules that need to be neutralized or destroyed before they wreak havoc with our biological processes. Molecules also loom large in diseases such as cancer—where the immune system is fooled by the body's own molecules gone bad—or Alzheimer's disease—where the culprits are protein molecules that misfold. Naturally occurring signaling molecules such as hormones are often implicated in diseases where their normal role as potent circulating modulators of tissue function is crippled or diverted. Nowhere is the argument for developing new technologies more compelling than for the purpose of identifying, characterizing, and altering molecules to improve our quality of life. The new medicine that is emerging, based on the molecular revolution described in this book, brings promises of cures that will truly exalt the molecule.

—Nadia Rosenthal, Ph.D.
Head of Outstation,
European Molecular Biology Laboratory,
Rome, Italy

Preface

In laboratories, clinics, and companies around the world, an amazing revolution is taking place in our understanding of life. It will dramatically change the way medicine is practiced and have other effects on nearly everyone alive today. This revolution makes the news nearly every day, but the headlines often seem mysterious and scary. Discoveries are being made at such a dizzying pace that even scientists, let alone the public, can barely keep up.

The six-volume Genetics and Evolution set aims to explain what is happening in biological research and put things into perspective for high-school students and the general public. The themes are the main fields of current research devoted to four volumes: *Evolution, The Molecules of Life, Genetic Engineering,* and *Developmental Biology.* A fifth volume is devoted to *Human Genetics,* and the sixth, *The Future of Genetics,* takes a look at how these sciences are likely to shape science and society in the future. The books aim to fill an important need by connecting the history of scientific ideas and methods to their impact on today's research. *Evolution,* for example, begins by explaining why a new theory of life was necessary in the 19th century. It goes on to show how the theory is helping create new animal models of human diseases and is shedding light on the genomes of humans, other animals, and plants.

Most of what is happening in the life sciences today can be traced back to a series of discoveries made in the mid-19th century. Evolution, cell biology, heredity, chemistry, embryology, and modern medicine were born during that era. At first these fields approached life from different points of view, using different methods. But they have steadily grown closer, and today they are all coming together in a view of life that stretches from single molecules to whole organisms, complex interactions between species, and the environment.

The meeting point of these traditions is the cell. Over the last 50 years biochemists have learned how DNA, RNA, and proteins carry out a complex dialogue with the environment to manage the cell's daily business and to build complex organisms. Medicine is also focusing on cells: Bacteria and viruses cause damage by invading cells and disrupting what is going on inside. Other diseases—such as cancer or Alzheimer's disease—arise from inherent defects in cells that we may soon learn to repair.

This is a change in orientation. Modern medicine arose when scientists learned to fight some of the worst infectious diseases with vaccines and drugs. This strategy has not worked with AIDS, malaria, and a range of other diseases because of their complexity and the way they infiltrate processes in cells. Curing such infectious diseases, cancer, and the health problems that arise from defective genes will require a new type of medicine based on a thorough understanding of how cells work and the development of new methods to manipulate what happens inside them.

Today's research is painting a picture of life that is much richer and more complex than anyone imagined just a few decades ago. Modern science has given us new insights into human nature that bring along a great many questions and many new responsibilities. Discoveries are being made at an amazing pace, but they usually concern tiny details of biochemistry or the functions of networks of molecules within cells that are hard to explain in headlines or short newspaper articles. So the communication gap between the worlds of research, schools, and the public is widening at the worst possible time. In the near future young people will be called on to make decisions—large political ones and very personal ones—about how science is practiced and how its findings are applied. Should there be limits on research into stem cells or other types of human cells? What kinds of diagnostic tests should be performed on embryos or children? How should information about a person's genes be used? How can privacy be protected in an age when everyone carries a readout of his or her personal genome on a memory card? These questions will be difficult to answer, and decisions should not be made without a good understanding of the issues.

I was largely unaware of this amazing scientific revolution until 12 years ago, when I was hired to create a public information office at one of the world's most renowned research laboratories. Since that time I have had the great privilege of working alongside some of today's greatest researchers, talking to them daily, writing about their work, and picking their brains about the world that today's science is creating. These books aim to share those experiences with the young people who will shape tomorrow's science and live in the world that it makes possible.

Acknowledgments

This book would not have been possible without the help of many people. First, I want to thank the dozens of scientists with whom I have worked over the past 12 years who have spent a great amount of time introducing me to the world of molecular biology. In particular I thank Volker Wiersdorff, Patricia Kahn, Eric Karsenti, Thomas Graf, Nadia Rosenthal, and Walter Birchmeier. Special thanks for a critical reading of the manuscript to Sandro Keller and the following members of his lab at the Leibnitz Institute for Molecular Pharmacology in Berlin: Anja Sieber, Nadin Jahnke, Natalie Bordag, Gerdi Hoelzl, Jana Broecker, and Sebastian Fiedler. My agent, Jodie Rhodes, was instrumental in planning and launching the project. Frank Darmstadt, executive editor, kept things on track and made great contributions to the quality of the text. Sincere thanks go as well to the production and art departments for their invaluable contributions. I am very grateful to Beth Oakes for locating the photographs for the entire set. Finally, I thank my family for all their support. That begins with my parents, Ed and Jo Hodge, who somehow figured out how to raise a young writer, and extends to my wife and children, who are still learning how to live with one.

Introduction

More than two centuries ago, a group of American colonists boldly declared their independence from a distant king at the same time that a young chemist in Paris started a much quieter revolution. In his laboratory next door to the king's Office of Gunpowder, Antoine-Laurent Lavoisier (1743–94) was performing a series of experiments that would eventually overturn the leading theory of life. His colleagues still believed that plants and animals were made of four basic elements—earth, air, water, and fire—which required a mysterious vital force to animate. In the 1770s that was the extent of people's knowledge of the chemical basis of life, and it was a scientific dead end.

Lavoisier discovered that air was composed of two substances: oxygen and nitrogen. Then he showed that oxygen could be recombined with another substance—hydrogen—to produce water. Suddenly he understood that most liquids, gases, and solids—including living tissues—were mixtures that could be separated into more basic components and analyzed. Lavoisier eventually lost his head to the guillotine when the French toppled their monarchy. But by that time he had helped usher in the modern age of chemistry, which would lay the groundwork for a new approach to the study of life. Within 200 years of Lavoisier's work, scientists had identified the main building blocks of life—DNA (deoxyribonucleic acid), RNA (ribonucleic acid), and proteins, lipids, and carbohydrates—and begun to understand the basic principles by which they work. Most of today's biology and biomedicine centers on studies of these molecules, and comprehending the nearly daily headlines about exciting developments in these fields requires a basic understanding of what they are and how they function. This book introduces the main molecules of life and how modern science investigates them. While the book touches on some general principles of chemistry and physics, it starts from the assumption that the reader has no background in these fields and is written for high school students and the general public.

The roles of DNA, RNA, and proteins finally began to become clear in the 1950s, and since then the pace of new discoveries has been very rapid. Today the entire genetic codes of humans and many other organisms have been read, and scientists are able to manipulate them by designing new genes or transferring them from one organism to another. These discoveries have ushered in a revolution that is beginning to have a tremendous impact on medicine and many other fields. Biology, chemistry, and physics are being drawn together into a new synthetic vision of life. Thanks to new technology, scientists are able to observe the dance of organic molecules as they work together in the cell, as parts of "molecular machines," and build tissues, organs, and whole organisms. Understanding how these levels of structure work together is not only answering fundamental questions about life; it also promises to help cure some of the major diseases that plague our species. Sometimes a person's entire existence is intimately linked to a single molecule. There are 3 billion letters in the genetic code in each of a person's cells, and thousands of diseases have now been traced to changes in single, key letters. Finding cures to genetic diseases will require a deep understanding of the mechanics of single molecules.

The Molecules of Life introduces how the chemistry and physics of organic molecules drive processes within cells and permit the construction of amazingly complex plants and animals. The first chapter describes some of the basic concepts needed to understand the field and discusses the methods used to explore the structures and behavior of DNA, RNA, and proteins. The rest of the book is devoted to stories about how these molecules carry out the business of life. Each chapter describes one of the major functions they perform: how the cell uses hereditary information stored in DNA; how signals are passed between cells and transmitted within them, and how molecules are delivered to specific locations, giving cells their shapes and forms. The final chapter is devoted to two of the most interesting areas of modern biomedical science: understanding how defects in molecules lead to disease and new molecular approaches to finding cures.

1

The Physics and Chemistry of Life: Basic Principles and Methods

An ancient, mystical Jewish tradition called Kabbalah teaches that the Bible holds all of the secrets of the universe, including power over life and death: If the words of the prophets could be read correctly, they could be used to perform miracles. According to a legend from the 16th century, the rabbi Judah Loew of Prague summoned this power to animate the golem, a man made of clay. Bringing the dead to life was also the theme of Mary Shelley's book *Frankenstein,* published in 1818, in which Victor Frankenstein assembles a monster out of the parts of dead bodies. Shelley did not describe how the body was brought to life, but in the introduction she claimed to have been inspired by the experiments of Luigi Galvani (1737–98), inventor of the battery. He had discovered that stimulating the nerves of dead frogs with electricity caused their legs to kick.

Today scientists know that the difference between inanimate and animate objects is not the result of incantations or electricity. Instead, life is made possible by interactions between organic molecules, chiefly *DNA* (deoxyribonucleic acid), *RNA* (ribonucleic acid), and *proteins.* These molecules behave in certain ways because of their structures, which means how their atoms fit together to make three-dimensional building blocks that combine into larger and larger structures, ultimately making a body. Very little was known about the structures or functions of these molecules until the 1950s. Then a series of major discoveries in chemistry and physics exposed their roles in cells, changing biology so dramatically that it needed a new name. In a speech given in 1938, Warren Weaver (1894–1978), director of natural sciences at the Rockefeller Foundation in New York, had already coined a term for it. Seeing that the focus of the life sciences was moving toward the fundamental chemical units within cells, he suggested the field should be called "molecular biology."

This chapter introduces the basic concepts and methods needed to understand how DNA, RNA, proteins, and other molecules work together to create living beings. The main discoveries that have been made about these molecules have come from a combination of methods from chemistry, physics, biology, and computer science. As such technologies continue to be created and improved, they are changing the way scientists see the structure and functions of molecules and their relationships to larger processes in organisms.

THE CENTRAL DOGMA OF MOLECULAR BIOLOGY

In 1953 a young American geneticist named James Watson (1928–) and his British colleague Francis Crick (1916–2004) had a brilliant insight into the building plan of the DNA molecule that accomplished several things:

- It demonstrated that DNA almost certainly contained the hereditary material of cells and organisms;
- It revealed how cells could copy DNA to pass it along to their offspring; and
- It showed how the molecule might change through mutations, which make evolution possible.

The new structure proposed by Watson and Crick caused a sensation in the scientific community because it made sense of some physical measurements that their British colleague Rosalind Franklin (1920–58) had made of DNA. It also explained studies that measured amounts of the four DNA building blocks, or *nucleotides,* in different species. The story is told in detail in chapter 2; it is mentioned here because it launched a new era in biology that shifted the focus toward the interactions between different types of molecules in cells.

Researchers had known of the existence of genes for half a century; the chemical substance they were made of, however, remained a mystery. The discovery of Watson and Crick immediately answered that question while raising many new ones. DNA not only was a library of information passed down from parent to child, or plant to seed, but also played an active role in the daily life of the cell. This was demonstrated in the 1940s by the American geneticists George Beadle (1903–89) and Edward Tatum (1909–75). Their experiments showed that a mutation in a single gene caused a type of mold to lose a single enzyme (a type of protein). This strongly suggested that whatever each gene was made of, it was responsible for the production of one protein. DNA and proteins were completely different kinds of molecules. How was the information in genes translated into another form? Beadle had already said that the issue would probably turn out to be very complex. In an article in a 1945 edition of *Physiological Reviews,* he stated that his work did not mean that genes directly make proteins. "In the synthesis of a single protein molecule," he wrote, "probably at least several hundred different genes contribute. But the final molecule corresponds to only one of them and this is the gene we visualize as being in primary control."

Crick provided a road map toward finding the answer in 1958 when he stated what is called the "central dogma" of molecular biology: "DNA makes RNA makes proteins." In other words, genetic information was transcribed into an intermediate molecule called RNA, which was then used to make a protein. No one knew how that happened; Crick's hypothesis was a challenge to the entire scientific community to figure it out. Within about 15 years researchers around the world had worked out the main steps in the process. Predictably, the answers raised

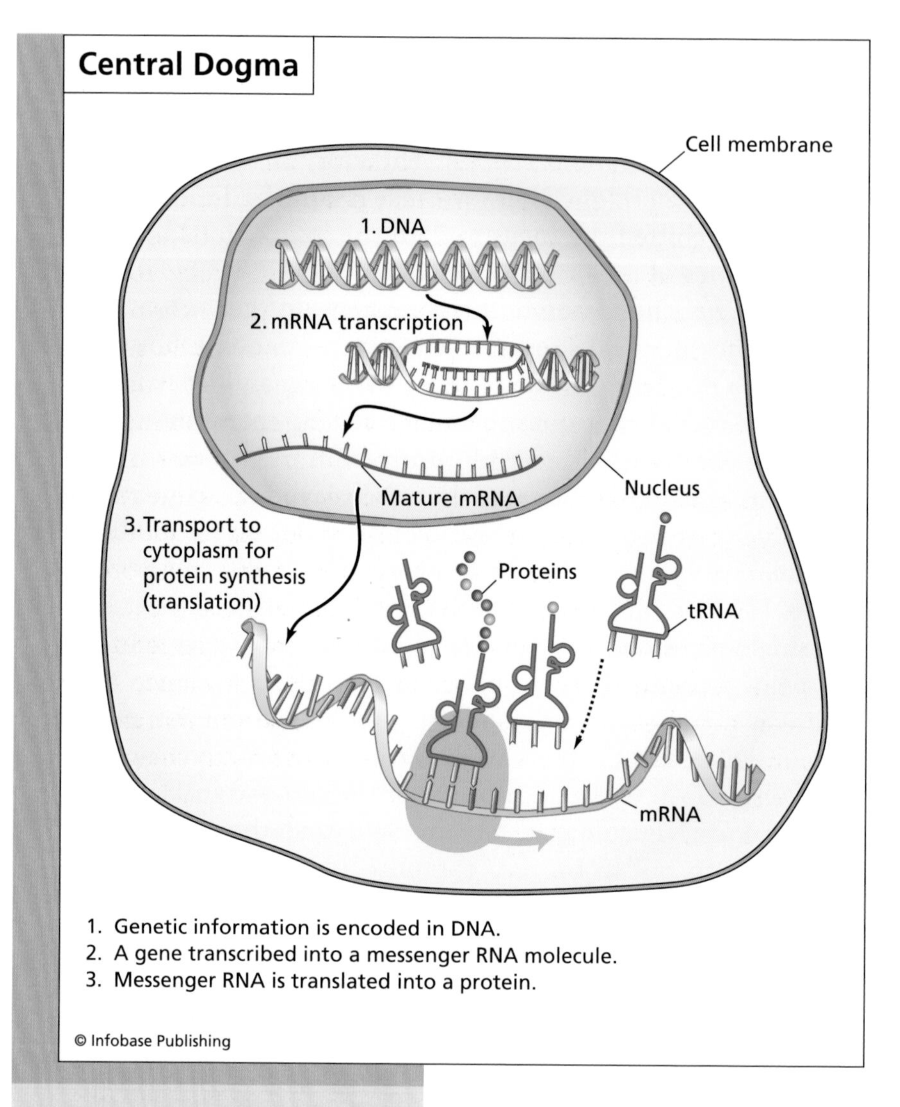

Francis Crick's statement "DNA makes RNA makes proteins" established a road map for research in molecular biology between the late 1950s and 1970s. Working out the details of the pathway between an organism's genome and the molecules it produces is still the major focus of many laboratories throughout the world.

new questions that in one way or another are the subject of most of the work going on in today's biological laboratories.

Crick's central dogma had several implications. The theory that "DNA makes

RNA makes proteins" meant that information was transmitted in a one-way direction. Each RNA was produced from the information in a gene, but the RNA could not send information back and change the gene. Likewise, an RNA could be used to make a protein, but a protein could not influence the content of the RNA. As chapter 2 will show, researchers have since discovered interesting exceptions to this rule; for example, RNAs can indeed sometimes rewrite the information of genes. Beadle and Tatum's principle of "one gene makes one enzyme" has also proven to be too simple. It is true that a gene cannot encode two completely different proteins, but one RNA can be used to produce very different forms of one protein, the way that the same ingredients and the same recipe can lead to different dishes if a cook leaves out steps, adds new ones, or changes their order. At every step in the transformation of genetic information into proteins, cells have evolved control mechanisms that enable them to step in and refine or block the process. The history of molecular biology since 1950 has been a process of working out the details and the exceptions to the central dogma, so it is one of the main themes of this volume.

FROM ATOMS TO MOLECULES

Most biologists think of what goes on in cells in terms of the activity of genes, RNAs, proteins, and other molecules. To understand how these molecules behave, it is necessary to go to a deeper level and examine their structures. The task is like that of an engineer. Building a copy of a machine or repairing it requires having a list of its parts. Each component has to have the right shape and size so that it can be combined with other components. And just as a motor has large parts that are built of smaller parts, molecules have different levels of structure, all of which are important to their functions. Proteins will be used as an example in this section to illustrate these levels of organization. The structures of DNA and RNA molecules are different and are covered in the second chapter, but the general principles needed to explain their behavior are similar.

Biological molecules are made of atoms that form bonds with each other. This is not a chemistry book, but it will be helpful to understand a little bit about the process of binding. Atoms have a central nucleus made of protons and neutrons, which is orbited by electrons. Early models of the atom depicted this as a sort of miniature solar system, but that changed in the early 20th century with the arrival of a new kind of physics called "quantum mechanics." In the new model of the atom, electrons have "fuzzy" positions that are located somewhere in orbital paths around the atom called "shells." There are several of these zones at various distances from the nucleus.

Each shell can hold a limited number of electrons. The innermost shell can hold two electrons, and each of the next two can hold eight. But often the shells are not filled; they lack one or more electrons. Atoms are not stable until their shells are filled; if their shells are not full, they try to fill them by binding to other atoms and sharing their electrons.

The simplest atom is hydrogen. It only has one electron, so its outer shell—which could hold two—is lacking one. Hydrogen easily forms bonds with other atoms that lack an electron in their own outer shells. Helium has two electrons and does not need to combine with other atoms. This makes it like neon, argon, and several other "noble gases," all of which have complete outer shells and are "inert" because they do not readily participate in chemical or biological reactions. This is much different from the behavior of oxygen, for example. Oxygen has eight electrons, split up in two shells: two in the inner shell and six in the outer one. This means that it lacks two electrons (to get a full set of eight in the outer shell). It can easily fill those spots by linking up with two hydrogen atoms, creating the molecule H_2O, or water.

Atoms that bind to each other by sharing electrons are linked in what is called a "covalent bond." Groups of atoms (molecules) can also link to each other by covalent bonds, but as a group grows larger, its overall shape plays an increasingly important role in what it can bind to. Atoms have to be brought close to each other to share electrons. Sometimes the bonds in a molecule make it inflexible, putting atoms that could bind to each other out of reach. The situation is a bit like trying to mount a metal bracket onto a wall with holes that have been predrilled. If the holes in the bracket do not line up with those in the wall, mounting it may be impossible.

The formation and breakup of chemical bonds is essential to every biological process and will be discussed throughout this book. There are other "noncovalent" ways for one atom to link to others, such as ionic bonds, which occur when one atom "steals" an electron from another, rather than sharing it, and van der Waals interactions ("nonchemical" forces between atoms and molecules). But in general, the structures of proteins, DNA, and RNA and the linkage of biological molecules are mainly due to covalent bonds. They often involve hydrogen atoms because of the high quantity of water in the body and the ease with which this element is integrated into molecules.

LEVELS OF PROTEIN STRUCTURE

The most fundamental form of each DNA, RNA, and protein is a stringlike chain of smaller molecules. The word *molecule* will be used throughout this book to talk about assemblies of atoms of various sizes. (There are more specialized names for molecules, such as *peptides* or *polypeptides,* but for the purposes of the book, *molecules* will do.)

In proteins the basic units are called *amino acids.* Evolution has produced 20 types of amino acids, which are assembled in the cell by other molecules or derived from food and then strung together. Every organism needs all 20 types of amino acid, but not every organism can make them all. Human and animal cells, for example, are unable to make the amino acid leucine, so people have to obtain it through their diet. (It is easy to obtain, by eating other proteins.) On the other hand, the bacterium that causes tuberculosis cannot obtain leucine externally and has to make its own. A group of researchers called the X-MTB consortium, based in Hamburg and Berlin, Germany, hope to use this fact to fight tuberculosis. They are currently trying to make drugs that will interfere with the bacterium's leucine-making machine. Such a drug might kill the bacterium without damaging the cells it has infected. According to Manfred Weiss, who heads the project at a station of the European Molecular Biology Laboratory in Hamburg, "A laboratory strain of tuberculosis that cannot manufacture leucine does not reproduce when it invades human cells. To reduce the potency of the bacteria, it may be enough to block a single step in the leucine synthesis machinery."

Amino acids are small clusters containing from about nine to 30 atoms. At the core of each type is a carbon atom, which binds to other atoms on four sides. In one direction there is always a hydrogen atom. On the second side there is a unit called an "amino group" (one atom of nitrogen bound to two atoms of hydrogen). On the third is a "carboxyl group," made of one atom of carbon, two of oxygen, and one of hydrogen. A group of atoms called the "side chain" forms on the fourth side; what is found there makes each amino acid unique.

A protein consists of a long string of amino acids connected to each other in a specific order, like the letters that make up this sentence. Amino acids bind to each other when the carboxyl group of one links to the amino group of the next. The process of binding releases a molecule of water and leads to the formation of a peptide bond. Adding water can make the amino acids release each other again. This could happen naturally, by itself, but it would take a long time—about a thousand years if the molecule were left alone in water. The process can be sped up by enzymes that break the bonds and thus free parts of the molecule to interact in new ways. By doing so, enzymes drive the chemistry of the cell.

The information in a gene shows the cell how to make a protein by telling it which amino acids to use and which order to put them in. After that, chemical bonds form between the amino acids to create many types of folds and structures, sometimes with the help of partner molecules. The following four different levels of structure have to be considered to understand how molecules work together:

- *Primary structure* is the string, the linear "spelling" of a protein, the list of amino acids in the order in which they are connected by peptide bonds.
- *Secondary structure* forms because the amino acids that lie near each other in the string chemically attract each other and create small folds. This usually produces one of two shapes, a coil called an *alpha helix* or a flat *beta strand.* Strands often lie next to each other and link to each other in larger beta sheets. Secondary structures are usually linked to each other by loose parts of the string known simply as "linkers."
- The surfaces of helices and coils may easily bind to other parts of the protein because their atoms also lack electrons. *Tertiary structure* arises when the secondary structures attract each other and fold into larger modules called *domains.* The result is to hide some amino acids in the interior and leave others exposed on the outside where they can interact with other domains or molecules. A protein may have several such modules, connected by loose, linking strands. This gives the protein a specific three-dimensional shape that plays an important role in its functions. These shapes are often tight and can only be changed by physical force or by a significant change in the domain's chemistry that breaks some bonds and allows others to form.
- The outer surface of a complete protein with all of its domains can bind to other molecules to create *quaternary structure.* Such "machines" composed of multiple proteins (or combinations of proteins, RNA, DNA, or other molecules) carry out most of the work in a cell.

As the primary amino acid string folds into these larger levels of structure, its shape and chemistry become so specific and complex that it can usually only bind to a few partners and sometimes only one. This keeps the wrong molecules from combining and is the main reason that tens of thousands of different molecules are able to organize themselves into the precise, long-lasting structures that make up a cell.

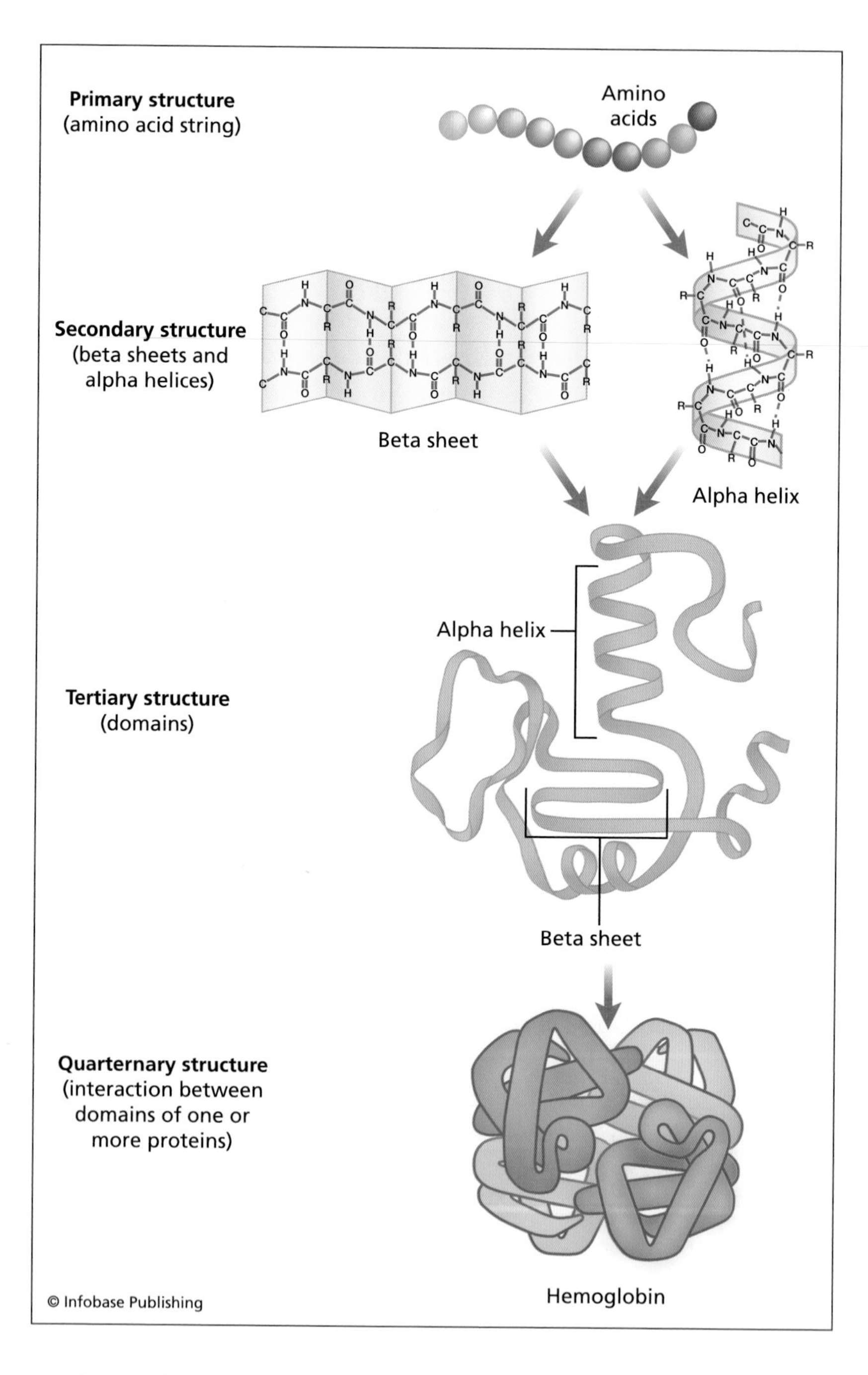

How these levels work together is exemplified in a protein machine known as an *exosome,* whose job is to break down RNA molecules that have become defective or are no longer needed.

(opposite page) A protein consists of a string of amino acids (primary structure). Chemical interactions between its subunits cause it to fold into small alpha helices or beta sheets (secondary structure), that form larger domains that determine the protein's functions (tertiary structure).

An exosome consists of 11 main proteins: It has a core made of six of the proteins, which dock onto each other to form a ring. The chemistry of this ring allows it to break the bonds between nucleotides, the subunits of an RNA molecule. The other proteins in the exosome help position an RNA and maneuver it into the core, putting it into contact with the ring proteins that break it down.

Some proteins bind to each other for long periods of time and make very stable machines. Others link only briefly to carry out a job; then the machine is taken apart again, and the pieces are reused to do other things. In 2002, researchers in Heidelberg, Germany, discovered how truly dynamic this situation is. Anne-Claude Gavin and Giulio Superti-Furga of the biotech company Cellzome worked with scientists at the European Molecular Biology Laboratory to capture the first complete view of the machines at work in a yeast cell. They found that 17,000 proteins form at least 232 machines. Many of them work in a "snap-on" way; they have a core of preassembled pieces, and when it comes time to do a certain job, a few more are added on. A machine may remain inactive until that happens. This gives the cell a way to control its activity. To be switched on, it may need to borrow the missing pieces from other protein complexes, or components may have to be made anew.

For example, one job of a protein called beta-catenin is to transmit instructions to several genes. When they are activated, the cell produces new molecules that can dramatically change its behavior. Because the instructions should only be transmitted at certain times, beta-catenin usually has to be kept switched off. The cell manages this by attaching other molecules to it. These molecules prevent beta-catenin from moving to the nucleus—which it has to do to activate genes—and they also call up other machines that break it down. When the cell's behavior needs to change, beta-catenin is attached to other molecules that deliver it to the nucleus.

HOW MUTATIONS CHANGE PROTEINS

Mutations in genes can be dangerous because they often cause changes in several levels of protein structure. The situation is a bit like the well-known parable from Benjamin Franklin: "For the want of a nail, the shoe was lost; for the want of a shoe, the horse was lost; and for the want of a horse, the rider was lost, being overtaken and slain by the enemy, all for the want of care about a horseshoe nail."

Altering one letter of DNA often leads to a change in one amino acid letter in the primary structure of a protein. This may change the close-range attractions between amino acids, which in turn can alter the formation of alpha helices and beta sheets (secondary structure), disrupting the shapes of domains (tertiary structure) and changing the set of other molecules that the protein can bind to. The effect is like putting a part designed for one type of car into the wrong model. If it does not fit, the part may break, possibly destroying the engine and even the whole automobile. A defective protein shape can break a molecular machine, and it may even kill the whole cell.

In the 1990s a number of studies by Peter T. Landsbury, a structural biologist at Harvard Medical School, Christopher Dobson of the University of Leeds, and other labs showed that different types of mutations can have similar effects on protein structure by transforming alpha helices into beta strands, then sheets, and finally tight clumps that the body cannot break down again. Such clumps were first found by the German physician Alois Alzheimer (1864–1915) in the brain of a woman who had died after suffering from a progressive disease that destroyed her mind. He found that dense *amyloid fibers*—made mostly of beta sheets—had collected between nerve cells in the brain, linked to the death of the cells and the symptoms of Alzheimer's disease. Since then fiber-forming proteins have been connected to about 20 different neurodegenerative diseases. These molecules and the diseases they cause will reappear as an important theme in chapter 5.

An entire field called *structural biology* is devoted to the study of the architecture of proteins and other molecules. One im-

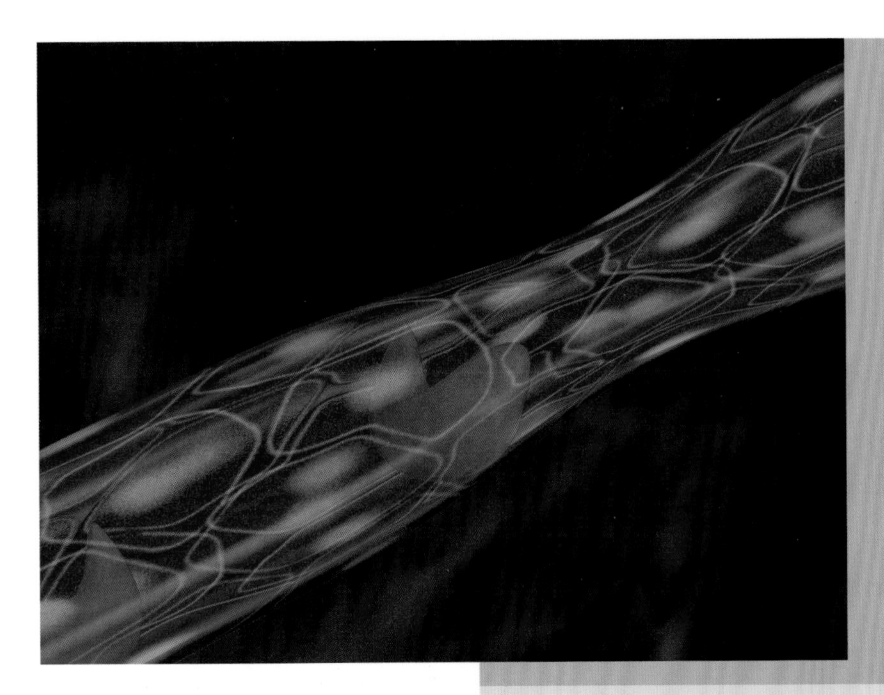

Mutations in the beta-globin protein cause hemoglobin proteins to form long fibers, distorting the shapes of red blood cells and disrupting the delivery of oxygen through the bloodstream. At the same time, the mutation offers people some protection from plasmodium, the organism that causes malaria, which grows and reproduces inside red blood cells. *(suttree.com)*

portant application of this work is the discovery and refinement of new types of drugs. The active ingredients of medicine are usually small molecules that work by docking onto proteins and changing their shapes and activities. For example, the surface of a protein may have a hole that has to be filled by a small molecule in order to be activated. If a drug plugs up the hole, that cannot happen.

This is why aspirin works. The drug had been used since 1899, when the Bayer company began to market it worldwide. But how it functioned was a mystery until 1971. That year John Robert Vane (1927–2004), a researcher at the Royal College of Surgeons in London, discovered that it prevents the production of fat molecules called prostaglandins by blocking sites in proteins called COX-1 and COX-2. The discovery earned Vane the 1982 Nobel Prize in physiology or medicine, along with the

Swedish biochemists Sune K. Bergström (1916–2004) and Bengt I. Samuelsson (1934–), for other key discoveries about the activity of prostaglandins.

By studying the map of a protein, structural biologists can discover the exact locations where molecules dock onto each other and determine what features a drug would have to have to interfere with a protein's activity. But to do so they need a three-dimensional structure of the molecule, which is difficult to obtain. As it is much easier to discover a molecule's sequence, structural biologists would like to learn to predict what secondary, tertiary, and quaternary structures a protein will form based on its amino acid spelling. Then it would be possible to plan and make artificial molecules shaped in a precise way, designed to control the behavior of other molecules. It might also be possible to see how a mutation would affect a protein's structure and behavior.

It is now possible to design small molecules with specific shapes, but scientists have not yet worked out the rules by which larger molecules achieve their folds. Some proteins contain thousands of amino acids, each of which is potentially able to bind to many others. The number of potential folds in a single molecule is so large that even today's largest computers are unable to calculate which ones will be produced by the chemical forces of all the amino acids. But it may be possible to find shortcuts, which is an area of intense research. In fact, every two years the U.S. National Institutes of Health sponsors an Internet competition based on this theme. Laboratories across the world are given a string of amino acids and use their computer programs to predict what shape it will form. The results have to be calculated purely by computer within two days. The winner in 2006 was Yang Zhang, an assistant professor at the University of Kansas. His entry was closest to the real shape—but still far from the structural biologist's dream of creating a detailed map of a complex molecule based on the sequence alone.

Protein structures are a good example of how strings of chemical subunits shape molecules and direct their behavior. DNA and RNA are made of different subunits, nucleotides, but the same basic principles of chemistry determine how they function. Their structures will be discussed in the next chapter.

BASIC MOLECULAR FUNCTIONS

Other chapters in this book look at how the structures of molecules permit them to carry out the processes that cells need to survive and build complex organisms. This section briefly introduces the relationship between DNA, RNA, and proteins and describes some of their basic functions. The easiest way to do so is probably to describe a day in the life of a typical cell.

Cells have to be flexible to survive, whether they live as single organisms in a pond or as parts of a vast human body. They need to move toward sources of food, to respond to changes in the environment, and to protect themselves from damaging substances (or predators) in the environment. These and a cell's other behaviors are carried out as a dialogue between the information in its genes and the environment. Its DNA sequence is like a huge library of user's manuals that has been in the writing since the beginning of life on Earth. The genome contains instructions that have helped the species survive over its entire evolutionary history. When circumstances change, the cell tries a different set of instructions, as if consulting a new user's manual. It shuts down some genes that it has been using and switches on another set, which changes its proteins and rebuilds molecular machines.

Changes in the cell are usually triggered by the environment. By "tasting" the molecules around it, the cell learns what to become. The tasting is done by proteins called *receptors* that float on the cell surface.

Foods have different, recognizable tastes because nerves in different parts of the tongue detect particular flavors, and all of these sensations are combined on their way to the brain. Environmental signals are also combined to have an effect on cell behavior. Cells have many types of receptor proteins, each able to detect a different "flavor" through a partner molecule called a *ligand.* Each type of cell produces a unique set of receptors, which enables it to taste some things but not others. Each environment tastes different because it contains a unique set of ligands. Muscles in the heart produce different ligands than do the linings of blood vessels, the bone marrow, or the brain. By interpreting combinations of signals, cells learn how to develop, when to divide, and even when to die.

Tasting is the first step in a process that tells a cell what to become and how to behave. Receptors send information into the cell via *signaling pathways.* These are information chains made of molecules that brush by each other, often binding for just an instant, just long enough to change each other's chemistry. Through signaling pathways particular molecules are triggered in a specific sequence, called a "cascade," which ends with the activation of certain genes. They work a bit like a telephone list that a teacher might use to get information to parents. Instead of calling every parent himself, the teacher calls one parent, who calls another, who calls the next parent on the list. Each bit of news reaches the same parents in the same order. Sometimes the lists are organized for speed: One person in the chain calls several. (If the teacher calls four parents, and each of them calls four more, a whole school of 1,000 pupils could be reached in just five steps.)

In the cell, information is passed from a receptor (teacher) to proteins (parents) using chemistry rather than telephone calls. When a signal triggers the same proteins in the same order it is called a pathway. Just as there are many receptors, there are many pathways. Signaling pathways cope with the same types of confusing situations that sometimes happen with telephone lists. One teacher may need to send out information to several classes. (Some receptors can trigger multiple pathways.) If there are three children in a family, a parent will be on three telephone lists. (One protein may belong to several pathways.) If a parent cannot reach the next person on the list, the chain may be interrupted (a broken or missing protein may block the signal). A child may use the list to make prank phone calls—just as cellular pathways can be abused. (Viruses often use receptor proteins, for example, to dock onto cells and gain entry through the membrane, much as if a telemarketer had found a copy of the telephone list.)

Eventually the information makes its way to the genes in the cell's large internal compartment, the nucleus—a bit like customer orders arriving at the kitchen of a restaurant. DNA contains the recipes for other molecules needed by the cell (RNAs and proteins). Information from signaling pathways switches on some genes and switches off others. This changes the set of molecules present in a cell, the way the arrival of new customers changes the food on restaurant tables. Tracking each order—identifying which receptor sends a certain signal, which proteins pass it along, and which genes are affected by the information—is the subject of a great deal of today's biological research.

Signaling pathways are involved in nearly everything that happens in the cell—good and bad. They can also break down. A protein may go missing, or it may get stuck in a "transmitting" mode where it broadcasts signals all the time, even when there has been no call from the receptor. If the pathway tells the cell to divide all the time, rather than alternating reproduction and resting phases, the result may be cancer.

One process that is controlled by signaling pathways is whether cells stick together in a tissue or migrate. The ability to switch on and off migratory behavior played an important role in the evolution of the first animals more than a billion years ago and is essential throughout the body of every animal alive today. When the signals that control this behavior break down, the result may be serious trouble, including cancer and metastases. An organism's body enters uncharted territory. Cancer resembles what happens all of the time in an embryo: Cells migrate to specific places to build new tissues. Tumors are like futile attempts by the body to build new organs that have no function and have not been shaped by evolution.

Signals trigger other types of molecular functions that are needed by cells. One is building the cell's architecture, which is crucial to its behavior. For example, the treelike structure of a neuron, with large networks of roots and branches, allows it to establish contact with other cells and communicate with them.

A red blood cell has to be small and flexible to navigate through tiny blood vessels called "capillaries." Muscles need pistonlike fibers to be able to expand and contract. Tubes and fibers made of protein complexes give cells these shapes.

A cell also needs to convert substances from the environment into what it needs to survive and make new copies of itself. Machines made of proteins and RNAs break down raw materials and glue them together in new ways, creating energy and nutrients. Another task is to protect the cell from dramatic changes in the environment and other dangers, such as invasions by viruses or bacteria.

Starting with chapter 2, *The Molecules of Life* shows how molecules work together to carry out these various tasks. The rest of chapter 1 is devoted to discussing some of the methods that have been developed to study molecules.

X-RAYS AND CRYSTALLOGRAPHY

Many of the instruments that have been built to study proteins and other molecules were developed for physics, as tools to investigate atoms or energy at work on the atomic scale. These inventions have proven so useful in the study of biology that many of their designers have received Nobel Prizes for their work. One of the most important developments has been the use of X-rays to investigate the structure of biological molecules.

For several decades after the German physicist Wilhelm Röntgen (1845–1923) discovered X-rays, doctors and others used them without being quite sure what they were. Some physicists thought that X-rays were made up of streams of particles; German physicist Max Laue (1879–1960) believed they were waves. Laue thought up an experiment involving crystals that could settle the issue. No one was quite sure what crystals were, either, although a popular hypothesis suggested that they were made up of rows upon rows of atoms, all arranged in the same direction. Individual atoms could not be seen using visible light because its wavelength was much too broad—like trying to play the piano wearing a baseball glove. Hitting single notes (seeing single atoms) would require thinner fingers. Laue reasoned that the wavelength of X-rays ought to be about the same as the distances between atoms in a crystal. This meant that shining them through the crystal ought to produce clean, single "notes"—the waves would be scattered in a pattern that could be interpreted.

The experiments revealed the wavelike behavior of X-rays and showed that crystals are indeed made up of row upon row of small repeated unit cells, each of which contains an identical package turned the same way. Imagine a huge stack of identical shoeboxes, each containing a pair of the same type of shoe, all the same size. The boxes could be stacked in many different ways, but since they are all identical, the shape of the entire stack will have some regular features. This is why crystals have well-organized geometrical shapes. A structural biologist hopes to obtain a three-dimensional image of a single protein, much like trying to determine the shape of a single shoe in one of the boxes.

When X-rays are aimed through a crystal, most of the energy passes straight through. But some of the rays collide with electrons in the atoms in the sample and are diffracted, or radiated off at an angle. These waves are captured on a photographic plate or a detector.

It is hard to describe and imagine this process because an X-ray diffraction experiment does not produce an image that makes sense to the eye, but the following analogy will give a sense of the problem and the solution. Suppose that pairs of identical shoes are taken out of their boxes and stacked on row after row of shelves. Imagine that each shoe is covered with sparkling sequins. If a huge light were shone through the shelves, it would sparkle off the sequins and cast light everywhere—off the walls, floor, and ceiling. If the shoes were all lined up identically—which is what happens to atoms in crystals—sequins on all the toes would reflect their light in the same direction, and so would the sequins on the sides and heels. Studying a *diffraction pattern* is like capturing all of the reflected light and analyzing it to determine the shape and relative positions of individual pairs of shoes.

To obtain a meaningful pattern, a molecule has to be organized in a very regular form, such as the identical unit cells of a crystal or the racks of a shoe store. Another arrangement that can be analyzed is a fiber if it is composed of a string of units that repeat over and over. DNA is such a fiber. In the late 1930s and 1940s William Astbury (1898–1961), a physicist working on biological molecules at the University of Leeds, stretched it and examined it using X-rays. The work was continued and refined in the early 1950s by Maurice Wilkins (1916–2004) and Rosalind Franklin, two physicists working on organic molecules in London. The experiments provided crucial information about the structure of DNA. They showed, for example, that it was arranged in the form of a helix; they revealed the width of the helix and the distance between individual "steps." These measurements were essential when Watson and Crick made their model of the double helix, described in the next chapter.

The first X-ray pictures of molecular structures came from salt and minerals, substances with a small number of atoms arranged in simple patterns. The British researchers Sir Lawrence Bragg (1890–1971) and his father, Sir William Bragg (1862–1942), shared the 1915 Nobel Prize in physics for this work. Their research laid the groundwork for the structural study of more complex molecules, which was carried out in new crystallography departments that were being set up at British universities, by young scientists who had studied in the Braggs' laboratories. One of them was Astbury. Before working on DNA, he used X-rays to examine protein fibers. His data allowed Linus Pauling (1901–94), an American chemist at the California Institute of Technology, to discover the basic secondary structures of proteins: alpha helices and beta sheets. It was a huge accomplishment that helped secure Pauling the Nobel Prize in chemistry in 1954.

John Bernal (1901–71), a member of William Bragg's laboratory in London, moved north to Cambridge University to start a crystallography unit there. One of his students, Dorothy Crowfoot Hodgkin (1910–94), became interested in the possibility of using X-rays to examine proteins when she was given a sample of the *hormone* molecule insulin, in crystal form. Bernal and Hodgkin took the first X-ray images of the crystal in 1934, but they had no way of interpreting their results into a model of insulin's structure. Yet, the protein crystals produced clear diffraction patterns, giving the researchers hope that they could someday be used to determine the arrangement of atoms in the molecules.

Another of Bernal's students, Max Perutz (1914–2002), had developed an interest in protein structure. Perutz had come to Cambridge in 1936 from Austria, where he had completed a degree in inorganic chemistry. He left the country when the Nazis invaded Austria and Czechoslovakia; his family became refugees. When war broke out between Britain and Germany, Perutz's nationality made him suspect, and he was briefly imprisoned. Upon his release he was put to work for the British war effort. He had done some work on ice crystals in glaciers and was assigned a crazy project to build artificial icebergs that could be used as aircraft carriers. (The plan did not work.)

Perutz had learned of the work of Pauling and other pioneers in the field during an organic chemistry course in Austria. In postwar Britain, finally free to pursue his own interests, he obtained crystals of the protein *hemoglobin,* which he hoped to use to obtain a structure. Bernal taught him how to handle the X-ray equipment, and Perutz began what would become a 22-year project to obtain a structural map of hemoglobin. In 1947 he was joined by a British student, John Kendrew (1917–97), who quickly became equally passionate about protein structure.

For many years it seemed as though the molecules would never give up the secrets of their architecture. Proteins are chemically very complex—with 20 amino acids, rather than a single atom like a diamond or the repeated structures of an artificial fiber. The patterns produced in X-ray experiments were so complicated that they did not seem possible to interpret. Additionally, the patterns were very weak. X-rays interact with an atom's electrons, so it is easiest to obtain diffraction patterns from heavy atoms with many electrons, such as metals. Biological molecules are composed almost entirely of light atoms with very few electrons. In addition, diffraction patterns did not capture phase, a type of information that was crucial to matching spots on the pattern to atoms in a molecule. Phase had to do with the fact that two X-ray waves, deflected by a protein crystal, might interfere with each other. Usually they cancel each other out. The X-rays that escape make the diffraction pattern. The spots on the photographic plates recorded the intensity of a wave but not its phase. It was like hearing all the notes of a piece of music played at the same time, without experiencing the sounds separated by time, or structured by a rhythm.

To solve these problems Perutz and Kendrew discovered they could soak protein crystals in heavier atoms, such as mercury, that would attach themselves to the side chains of some of the proteins. These atoms gave off strong, characteristic signals that could be used to locate and chart the positions of other atoms. They could also be used to calculate the phase, by comparing the patterns made by proteins soaked in the atoms with an image from unsoaked crystals.

Producing the first structure of a protein, myoglobin, took five years of work on the part of Kendrew and a team of women "computers" to do calculations by hand, figuring out the mathematics of 250,000 spots on a plate of film. Perutz followed a similar procedure with hemoglobin. As the project progressed, they were able to make use of early computers, which was essential because about 5 billion sums had to be calculated based on the diffraction patterns. Finally, in 1959 the groups had built three-dimensional models of the two proteins at a resolution of two angstroms (an angstrom equals one ten-billionth of a meter), insufficient to detect the lightest atoms, such as hydrogen, but clear enough to reveal the main shapes within the molecule. For their work, the two men were awarded the 1962 Nobel Prize in chemistry. That same year two more of Perutz's former students, Watson and Crick, were presented with the Nobel Prize in physiology or medicine for their work on DNA.

The Nobel ceremonies include a lecture in which the awardees explain their work. In his presentation Perutz noted that obtaining the structure was one step toward a larger goal: explaining how the protein's building plan allowed it to transport oxygen. He said: "We may hope that the interaction, and the acid shift on which the respiratory functions of hemoglobin depend, will eventually find their explanation in terms of the structural changes of which these new results have just given us a first glimpse." To do so would probably require obtaining an even higher-resolution image of at least one of the two forms, he stated: "Due to the enormous amount of labour involved this may take some time, but not much, perhaps, compared to the 22 years needed for the initial analysis." In fact, just a year after receiving the prize, Perutz was able to report success. A clearer set of data and a new crystal revealed that some of the internal structures of the protein rearranged themselves when oxygen atoms were bound to the molecule. The two different conformations of hemoglobin were like snapshots; Perutz could now assemble them into a "film" showing how the protein's structure changed to carry out its functions.

Protein crystallography has now become the main method used to unravel the structures of proteins. What took Perutz and Kendrew more than two decades can now be done in a single afternoon—once scientists have a usable crystal—thanks to new technology and analytical software. The small X-ray machines used in laboratories in the 1950s and 1960s have been replaced by some of the world's largest machines: *synchrotrons,* ring-shaped tunnels that are used to smash atomic particles together. In 1969 the British researcher Ken Holmes (1934–), who had studied with Franklin, carried out a series of experiments on a synchrotron in Hamburg, Germany, showing that the high-energy X-rays produced by these machines could be used to obtain protein structures. He discusses the motivations behind his pioneering experiments and some of the results in the sidebar.

NUCLEAR MAGNETIC RESONANCE

Nuclear magnetic resonance (NMR) is a second method adapted from physics that is now commonly used to reveal the structures of proteins and other molecules. It does not require proteins in crystal form, which means that it overcomes some of the problems faced by crystallographers.

The First Synchrotron Experiments

Ken Holmes was a Ph.D. student in Rosalind Franklin's laboratory in London from 1955 to 1959. After spending two years as a postdoctoral student at the Children's Hospital Boston, he returned to the Laboratory of Molecular Biology (famous for its scientists having accrued 13 Nobel Prizes) in Cambridge, England. In 1968 he became a director at the Max Planck Institute for Medical Research (recipient of four Nobel Prizes) in Heidelberg, Germany. In the late 1960s Holmes and his colleague Gerd Rosenbaum carried out the first experiments on protein structures using X-rays produced by the electron synchrotron DESY, a huge ring-shaped particle accelerator used in physics experiments. Today electron synchrotrons or storage rings have become the main source of X-rays for diffraction experiments to obtain the structures of proteins and other molecules. In the following personal interview with the author, he shares some memories of those pioneering experiments.

Ken Holmes is former head of the Max Planck Institute for Medical Research in Heidelberg, Germany, and a pioneer in the use of synchrotron radiation for the study of biological molecules. In the 1950s he was a doctoral student under Rosalind Franklin. *(Russ Hodge)*

While I was in Cambridge, I became interested in trying to get better sources of X-rays. At the time the method people were using was essentially the same as that discovered by [Wilhelm] Röntgen, which involved creating a beam of electrons at moderately high energy [in a vacuum] and shooting them at a metal, in our case a chunk of copper, known as the anode. Why does this produce X-rays? The beam knocks electrons out of their deep shell in the copper atoms and as they return to that shell they give off fluorescent energy in the form of X-rays. However, the electron beam carries quite a bit of energy and so the beam easily burns a hole in the metal, or melts it. To stop this [from] happening you have to cool the metal, and rotate it [the anodes are often rotating metal discs]. Cambridge had a very good workshop that was making rotating anode X-ray tubes [called "tubes" because Röntgen's original apparatus was in a long glass tube], but their source of electrons—a sort of gun that shoots them at the metal—was not much good. So with my colleague Bill Longley we developed a much better electron gun and connected it with their anode. The head of the workshop had absolutely no faith in the whole thing and only grudgingly gave us one of their beautiful anodes. Nevertheless, the new X-ray tube worked, and we successfully marketed it. The intensity of the X-rays was much better but still far short of what we wanted. To get more intensity you need more electrons, but this produces more heat. To dissipate the heat you spin the disc faster. However, you can only spin a disk of metal

(continues)

(continued)

so fast before it flies apart. Moreover, you've got to cool the system somehow. One quickly reaches physical limits. Therefore, I began looking around for even better sources of X-rays.

At that stage the German government and University of Hamburg were planning to build a very powerful electron accelerator facility called the Deutsches Electronen Synchrotron (DESY). This is a huge ring-shaped tunnel containing an evacuated stainless steel tube in which electrons could be accelerated. This also begins with an electron gun, but instead of firing a beam of electrons into metal, it fires electrons into the ring. They are already traveling at about the speed of light, but then they are accelerated even more, and as that happens, they get heavier. When they reach their maximum energy they are deflected into a target, where they produce a menagerie of subatomic particles. Electromagnets mounted along the tunnel keep the electrons moving in a circle, rather than a straight line, but curving their trajectory causes the release of energy—intense radiation known as synchrotron radiation covering all wavelengths from infrared to X-rays—that fly off at a tangent at every bending magnet. This is a curse for the high-energy physicists (it increases their electricity bill enormously) but a blessing for us. If you insert a suitable window you can let the radiation out of the ring. One can then select out the X-rays with a suitable crystal monochromator. It turns out, you end up with an intense X-ray source. The X-rays then need to be focused into a tiny beam and directed at the biological sample, often a crystal, but our work involved viruses and muscle fibers.

In 1963 I wrote to DESY and suggested that we ought to use the ring as an X-ray source. I got an encouraging response, but at the time I was still in Cambridge, England, and it was not easy to do a project across international borders. Then I received the appointment in Heidelberg. One of my first students was a physicist named Gerd Rosenbaum, who had just finished his undergraduate studies at DESY. He had done experiments there and knew the whole set-up. He went back to Hamburg and began setting up our equipment. When the time came for the experiments, we were under intense time pressure. We only had 16 hours to mount our equipment, take measurements, and put everything back. We had to use an existing lab—physicists at the synchrotron had already set up an experimental laboratory to utilize the synchrotron radiation, but their interest was in ultraviolet radiation, not X-rays, so their setup was quite different from what we needed. They warned us: "You cannot perturb anything here; if you move anything, you have to put it back just the way it was." Somehow we managed, and in those 16 hours we got the first X-ray diffraction patterns ever taken with synchrotron radiation, actually with a bit of muscle fiber.

Even the synchrotron was not an ideal source of X-rays. The electromagnets in the ring are part of tuned oscillating circuits; in essence they are big inductive coils with condensers attached to them, and their magnetic field goes up and down 50 times a second in a synchronized way to match the ever increasing mass of the circulating electrons [hence the name *synchrotron*]. What this means for our experiments is that the electron beam in the ring is produced and killed 50 times a second, which is not

(continues)

(continued)

highly efficient! And it only gives off X-rays when it is at the peak of its energy, which is only 10 percent of the total time it operates. But DESY was already planning to create a "storage ring," where a steady beam of electrons at one energy would be kept in a constant orbit. If we could build an experimental station on that ring, we would have a much better continuous source of X-rays that could be used on a permanent basis for biological experiments. We built that station as a service facility of the European Molecular Biology Laboratory; over the last 30 years it has been used to obtain structures by protein crystallography of thousands of biological macromolecules by scientists from all over Europe.

Synchrotron radiation has become very important. Nowadays there are a couple of dozen electron storage rings distributed round the world that have been custom built as intense X-ray sources. Although the source is nowadays a storage ring rather than a synchrotron, the radiation is still known as synchrotron radiation. These sources have yielded data for the solution of tens of thousands of atomic structures of biological macromolecules, some, such as the ribosome, of immense size and complexity. The growth of protein structure determination by X-ray diffraction from an important but esoteric method to one of the pillars of modern molecular biology would not have been possible without synchrotron radiation.

Note: In a second interview in chapter 2, Holmes recounts some of his memories of working with Franklin in Cambridge in the late 1950s.

To build a crystal, proteins have to be captured in a liquid at high concentrations and then slowly dried until they lock onto each other in symmetrical arrangements. This method cannot be used to study all protein structures or to answer all of scientists' questions about them. First, not all proteins form crystals of a high enough quality to be studied. Second, crystals are often highly artificial forms of molecules. A protein's natural environment is liquid, where it moves in a flexible way as it finds and binds to partners. That flexibility is mostly lost in the rigid structure of a crystal. NMR avoids some of these problems because it can look at proteins in liquids or even a solid form.

The principles behind NMR were discovered in the 1930s by Isidor Rabi (1898–1998), a physicist at Columbia University. He was using molecular beams to investigate the forces that hold electrons to the nuclei of atoms, work that earned him the 1944 Nobel Prize in physics. It also had practical applications in the development of radar, a project that Rabi was recruited to work on in World War II. Two physicists who had been likewise recruited, Felix Bloch (1905–83) and Edward Purcell (1912–97), adapted NMR so that it could be used to investigate liquids and solid objects. Their work led to the 1952 Nobel Prize in physics. NMR slowly became an important method for the determination of structures of biological molecules thanks to the efforts of Kurt Wüthrich (1938–), a Swiss chemist who now heads laboratories at the Swiss Federal Institute of Technology in Zurich and the Scripps Institute in La Jolla, California. Wüthrich began working with NMR when he joined Bell Laboratories in Murray Hill, New Jersey, in 1967, where he had access to one of the best instruments in existence at that time. He used it to study the structure and behavior of proteins in liquids; thanks largely to his efforts, it has become a standard tool in structural biology, drug discovery, and related fields. His accomplishments were recognized with the award of the 2002 Nobel Prize in chemistry.

Obtaining a molecular structure requires finding positions of single atoms and establishing their positions relative to other atoms in the same molecule. NMR works by both identifying atoms and "interrogating" them about others that are nearby. It applies a very strong magnetic field to a liquid containing the protein (and sometimes other types of molecules) that a scientist wants to study. When placed in a magnetic field of a precise strength, the nuclei of some atoms absorb an amount of energy that can be precisely measured. Different strengths of fields are required to make different types of atoms behave this way, so energy measurements can reveal the presence of specific atoms. Finding their locations relative to each other requires relaxing the magnetic field again.

This effect is like moving a strong magnet next to a compass and then removing it. When it is close, the magnetic field draws the needle of the compass. Removing the magnet again makes the needle return to its normal position. The same effect happens with the nuclei of atoms in an NMR experiment. The magnetic field aligns them; when it is relaxed, they return to their normal state, but the way that they do so depends on what other atoms are nearby. Thus, the properties of one atom produce a "signature" that reveals what other atoms are nearby, and this can be used to reconstruct a map of a protein or another molecule.

NMR can only provide information about small regions of molecules. This limitation means that it usually does not provide the structure of an entire molecule. On the other hand, it can examine proteins in liquids, which is close to their natural state. This means that researchers can also observe changes in its structure, during processes such as the binding of a small molecule to a protein. That gives a look, for example, at how a drug interacts with its target.

The same principles and technology underlie magnetic resonance imaging (MRI), a medical technique. Instead of detecting mineralized substances such as bone, the way an X-ray machine functions, MRI detects the presence of particular fluids. This allows it to be used to study the structure and activity of the brain and other organs.

ATOMIC FORCE MICROSCOPY

The 1980s and 1990s saw the development of new methods to investigate the physical characteristics of proteins and other mol-

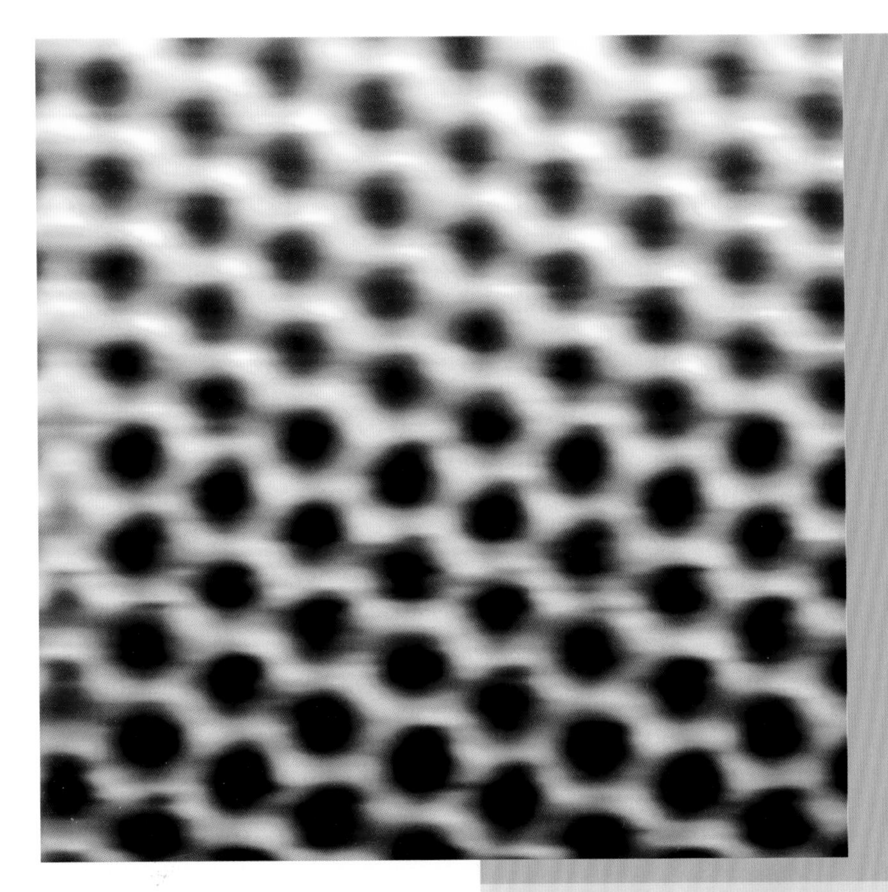

Atomic force microscopy was first used to investigate materials and surfaces. This study of graphite revealed all the atoms within the hexagonal unit cells, including a hidden atom. The image covers an area of 2 × 2 nm^2.

ecules. One of these, atomic force microscopy (AFM), studies the shapes and physical characteristics of molecules, cells, and surfaces by directly touching them.

In the early 1980s Gerd Binning (1947–) and Heinrich Rohrer (1933–), working at IBM in Zurich, developed an instrument that uses a tiny needle to feel its way along surfaces. The technology works a bit like an old vinyl record player, which has a very fine needle attached to an arm. The tip of this needle travels along the surface of a record and is jostled by its grooved surface. The machine turns these motions into sound. In an atomic force microscope, the needle is much tinier—its tip is a single atom in size. As it is moved, it does not actually touch down on the surface; instead, it is brought just close enough to interact with the atoms there. At very close range electrons from the atoms on the surface jump and cause the tip of the needle to move. The motion can be captured by tracking the movement of the arm through sensors or a laser. A computer assembles the information into an atom-by-atom image of the surface. This practical application of the jumping behavior of electrons earned Binnig and Rohrer the 1986 Nobel Prize in physics.

The instrument provides much higher resolution than light or electron microscopes. It has been used to make detailed maps of the surfaces of neurons and other cells, as well as to probe DNA and protein complexes. In early 2008 Daniel Müller's biophysics group at the University of Technology in Dresden, Germany, used the method to study cells that had been infected by *prions*. Misfolded prion proteins cause scrapie, mad cow disease, and several other diseases of the nervous system. In the test tube, prions assemble into tough fibers that do not get broken down, but this process had not been directly observed in cell cultures or in animals. Müller's group used AFM to probe the surface of mouse brain cells that had been infected. They found that long, prion-containing fibers began to grow on the cells, confirming that the behavior of prions in the test tube also happened in cells.

Another use of AFM has been to study the physical forces at work inside protein structures and how they are changed by external forces. Some molecules need to stretch or undergo other major changes of form to carry out their functions. The protein titin, for example, helps muscles to contract and relax. Muscle fibers are built of subunits called "sarcomeres," which have a pistonlike structure. A sheath of fibers moves along a much thicker fiber to slide the piston in and out. Titin is a long, stringlike molecule that holds the pieces of the sarcomere together and allows them to expand and contract. Researchers thought it could do this because it is composed of more than 200 tiny folded Ig modules (so-named because they closely resemble the structure of immunoglobulins, or antibodies), which act as springs. When titin has to stretch, physical force breaks some of the chemical bonds in the modules, and they unfold. When the molecule relaxes again, the bonds are reestablished, and the modules refold themselves. An additional region of titin, called "PEVK," is more stringlike and was also thought to act as a spring, but the process had not been observed directly, so it was impossible to tell which regions were doing the most work.

In 1997 Hermann Gaub and his laboratory at the Institute for Applied Physics in Munich, Germany, decided to use AFM to investigate this question. They attached one end of a titin molecule to a gold plate and grabbed the other end with the tip of the atomic force microscope. As the tip pulled, Gaub's team made two measurements: the amount of force that was needed to move the tip and the length of the molecule. The result was a sawtooth-like pattern in which the molecule suddenly became longer, then the force relaxed. As pulling continued, the amount of force rose again; the titin molecule became a little longer, and then there was another drop in force. This happened over and over. Gaub thought this represented the unfolding of single Ig domains, but it was hard to be sure. Each of the 200 Ig modules is encoded by a different part of the titin gene and had evolved slightly different features over the course of evolution. Even tiny differences meant that each domain behaved slightly differently, requiring a different amount of force to unfold. The scientists came up with a solution: They created an artificial molecule in which some of the Ig domains were replaced by identical copies. Now when they pulled, they saw several identical peaks—the same as the number of modules they had added. The experiments proved that AFM could be used to obtain new types of information about the structures and mechanical forces that govern the behavior of molecules.

PHOTONIC FORCE MICROSCOPY

The first photonic force microscope was developed in the late 1990s by the laboratories of Ernst Stelzer and Heinrich Hörber at the European Molecular Biology Laboratory in Heidelberg. The basic principle is similar to that of AFM: a tiny probe is used to investigate surfaces, molecules, and forces at the atomic scale. In this case the probe is a tiny plastic ball rather than a needle, and instead of being mounted on an arm, it is held in place by a laser beam.

The invention of photonic force microscopy (PFM) has its roots in a series of experiments carried out in the 1970s at Bell Laboratories in New Jersey. Researcher Arthur Ashkin began developing a new way of using lasers to manipulate nanoscale objects. He hoped to be able to control single atoms, which would have important uses in industry. The idea was to catch a particle in an optical "trap," a beam of light from which it could not escape. After years of work, a project at Bell Laboratories headed by Stephen Chu (1948–) successfully applied the technique to manipulate and cool atoms. The accomplishment earned Chu the 1997 Nobel Prize in physics.

By the mid-1990s Stelzer and his colleagues were using the method to create a new type of microscope. They could trap a tiny plastic bead in a laser and hold it still, then watch as nearby molecules affected its behavior. By coating the bead with *antibodies,* they could attach it to other proteins and watch how it moved. The bead's motion could be recorded very precisely, and this could be used to reconstruct the motions of proteins and the forces that acted upon them. An analogy is to think of a baseball pitcher throwing a glowing ball in the dark. With a film that tracked the path of the ball through the windup and pitch and a knowledge of human anatomy, a scientist could reconstruct the motion of the pitcher's arm.

One of the first experiments the scientists performed aimed to help resolve a controversy about cell membranes. These "envelopes" around cells are made up mostly of fat molecules called "lipids," which form a double layer—like one soap bubble enclosed in another, with just a tiny bit of space between the two. Floating in the layers are such proteins as receptors, which receive signals from the environment. Until the 1990s most scientists believed that the proteins and lipids floated more or less freely, independent of each other. Then Kai Simons, director of the Max Planck Institute of Molecular Cell Biology and Genetics in Dresden, and Elina Ikonen of the National Public Health Institute of Finland proposed a different model. They pointed out that proteins are glued to lipids deep within the cell. As they are transported to the membrane, some lipids and proteins remain linked to each other. Simons and Ikonen suggested that cholesterol and other lipids were being used as platforms—like small, floating rafts—on which several proteins could be assembled. This could allow the cell to preassemble groups of molecules that needed to work together. The hypothesis also helped explain how cells are able to build surfaces with different characteristics. Skin is a good example; one side is exposed to the environment, and the other side comes in contact with the inside of the body. Because these two surfaces have different functions, they need different proteins and lipids.

Hörber and Stelzer realized that PFM could be used to test the hypothesis. Scientists in their groups attached the bead to a protein in the membrane, thought to be embedded in a raft, and tracked its motion using the photonic force microscope. Then they repeated the experiment using nonraft proteins. The bead behaved differently in the two situations, a bit like a fishing lure: It is harder to drag a lure that has become clogged with moss than one that is free. By watching the motion of the bead, Hörber and his colleagues could measure the viscosity of the membrane, its "stickiness." They saw that raft proteins were not moving as freely in the membrane because they were attached to lipids. Another experiment that depleted cells of cholesterol—the glue that holds rafts together—released the proteins and allowed them to move much more freely.

PFM has also been used to study motor proteins, which are like the truck drivers of the cell. The cell is criss-crossed by an immense highway of fibers called *microtubules*. Microtubules and motors help build and change the architecture of the cell and play a key role in cell division and other processes. They are covered in more detail in chapter 4. Kinesin and other motor proteins travel along the surface of the microtubules. The motors have two "feet" that walk down the surface of the microtubules, attached to a towing line and then to the cargo. One question about motors has concerned the flexibility of the line: Does it have joints, like an arm? Which parts are stiff, and which are flexible? Ernst-Ludwig Florin, professor at the University of Texas at Austin and a former member of Hörber's group, used PFM to find out. He attached a bead to one of the motor's cargoes and filmed its motion at an incredibly high rate of speed—several hundred thousand frames per second. From the motion of the bead he concluded that kinesin is stiff when it moves, acting more like a towing bar than a rope.

USING LIGHT MICROSCOPES TO INVESTIGATE MOLECULES

The light microscope is experiencing a renaissance with the development of methods that have given scientists new ways to watch the behavior and activity of molecules. This is happening even though there has not been any significant change in the resolution of light microscopes since the early 19th century, when Joseph Lister (1786–1869) made some important changes in their lenses and construction. Under the best conditions a light microscope can magnify objects about 2,000 times. Even if the microscope could magnify things 2 million times—which is the resolution of the electron microscope—it would not be able to pick out proteins or other single molecules. Even so, light microscopy is undergoing a revival in the molecular age because of the discovery of fluorescent proteins and ways that they can be used to study living processes.

In the early 1960s Osamu Shimomura and his colleagues at Princeton University in New Jersey isolated two luminescent proteins from *Aequorea victoria,* a species of jellyfish. Doing so required Shimomura to undertake a seven-day drive from New Jersey to Friday Harbor Laboratories of the University of Washington in Washington State, then capture and dissect 10,000 jellyfish in just a few months every summer. Shimomura and his colleagues managed with the help of schoolgirls and the children of scientists. By the mid-1970s the researchers had perfected their routine and were collecting more than 3,000 jellyfish every day. There was heavy interest in these and other molecules that could transform chemical energy into light, such as the luciferase proteins found in fireflies.

Huge numbers of animals were necessary because jellyfish only produced the two fluorescent proteins in small amounts, and the researchers needed a great deal in very pure form. Material from 50,000 jellyfish had to be passed through 60 steps of purification, leaving at the end only about 200 mg of aequorin, one of the fluorescent proteins. Finally, in 1986 a laboratory in Georgia and another in Japan independently cloned its gene: They isolated the DNA sequence encoding aequorin. This meant that they could begin doing genetic engineering experiments with it.

In 1992 Douglas Prasher and colleagues at the Woods Hole Oceanographic Institution in Massachusetts cloned the green fluorescent protein (GFP), the second of the jellyfish fluorescent proteins. GFP absorbed blue light emitted by aequorin and flashed brilliant green. Prasher wanted to continue working with the molecule, but the funding for his project ran out. In the meantime, however, he had been talking about the molecule at conferences; in the public was Martin Chalfie, a scientist from Columbia University in New York. Chalfie immediately realized that it might be possible to turn the protein into a tool for microscopy. A laser microscope could play the role of aequorin; shining the right wavelength of light on GFP might trigger it to flash. The real use of the tool would come from the fact that the light-emitting part of the protein was contained in one small module, called a "domain." It might be possible to attach the domain to other proteins. If so, it would serve as a beacon that would allow the molecule to be tracked under the light microscope.

Several steps were necessary to turn GFP into such a tool. Prasher sent samples of the gene to Chalfie, whose laboratory introduced it into bacteria and frog cells. In "Green Fluorescent Protein as a Marker for Gene Expression," published in the February 11, 1994, issue of the journal *Science,* Chalfie and Prasher reported success: Both types of cells were capable of producing glowing GFP proteins. It had not been clear that this would work. Chalfie had been concerned that other jellyfish molecules might be needed to build or switch on the molecule. That was not the case, but there were other problems. To be seen in a cell, GFP had to be excited by the laser in a microscope, but it was not bright enough to give a clear signal. Furthermore, it did not work well at body temperature, which meant that scientists would have trouble inserting it into mammal cells.

The solution to these problems might be to rebuild the protein, but a close look at its structure was needed to determine what to change. S. James Remington, a structural biologist at the University of Oregon, obtained crystals of GFP and used X-rays to obtain a high-resolution map of the molecule. His lab discovered that GFP contains a unique structure not found in other proteins, a barrel-shaped domain made of 11 beta strands and one alpha helix. Three amino acids inside the helix are responsible for absorbing and emitting light. This structure suggested ways of improving the protein to one of Remington's collaborators, Roger Tsien, and his colleagues at the University of California, San Diego. Changing one of the three amino acids altered the wavelength of light needed to activate the protein so that standard lasers used in microscopes could do the job. Additional changes made the module much brighter and allowed GFP to work efficiently at body temperature. Since the late 1990s Tsien's laboratory and others have developed versions of the molecule that are activated in different ways and emit other colors, including blue, cyan (green-blue), and yellow. More fluorescent tools have been produced using proteins from coral and other animals. Through genetic engineering the proteins have become thousands of times brighter than their natural counterparts. Their immense usefulness for biology led to the award of the 2008 Nobel Prize in chemistry to Shimomura, Chalfie, and Tsien.

The fluorescence module of GFP can be attached to other genes, often leading cells to create versions of proteins that are fluorescent but otherwise normal. The altered protein emits light if it is illuminated by a laser microscope. This has several important uses. The first is simply to see if certain types of cells make a particular molecule. Another application is to discover where a protein works in the cell. Scientists still lack this information for the thousands of human molecules discovered during the Human Genome Project.

Another use of fluorescence microscopy is to find out whether two proteins work together, as, for example, when a protein is needed to activate another gene. This can be investigated by marking the two molecules with different colors and watching to see if the appearance of one protein is followed by the production of the other.

Some of the more elaborate uses of GFP and similar proteins are based on the physics of how they absorb energy from one type of light and then emit it as another. Each protein that is made can only do this one time. This means that activating a molecule with the laser of the microscope illuminates it, but then bleaches it. This fact has been turned into a method to study how the cell makes, transports, and recycles proteins. If a scientist wants to know how often a cell "reloads" its membrane with a certain protein, for example, the laser can be trained on the membrane and left there until a particular molecule has been bleached. If a fluorescent version of the molecule reappears at the membrane, it is because the cell is actively making new copies of it and transporting them there.

This method, fluorescence recovery after photobleaching (FRAP), allowed scientists their first direct look at how often the cell produces and replaces receptors. Philippe Bastiaens, who is director of the Department of Systemic Cell Biology at the Max Planck Institute of Molecular Physiology in Dortmund, Germany, has used FRAP to demonstrate that some types of receptor proteins are reused. His lab showed that after activation on the cell membrane, a receptor called Ras detaches itself, is carried into the cell, and is "reloaded."

Each GFP-tagged protein gives off a particular signal with precise characteristics. The signal shifts whenever the protein's activity changes, for example, when a signaling molecule has been chemically activated. This permits several other types of experiments; for example, scientists can measure whether two molecules bind to each other—important in the testing of potential new drugs. Most drugs are small substances that bind to an important protein and change its activity. The target protein can be labeled with a fluorescent marker. The drug candidate can be introduced into the cell, and then scientists watch for changes

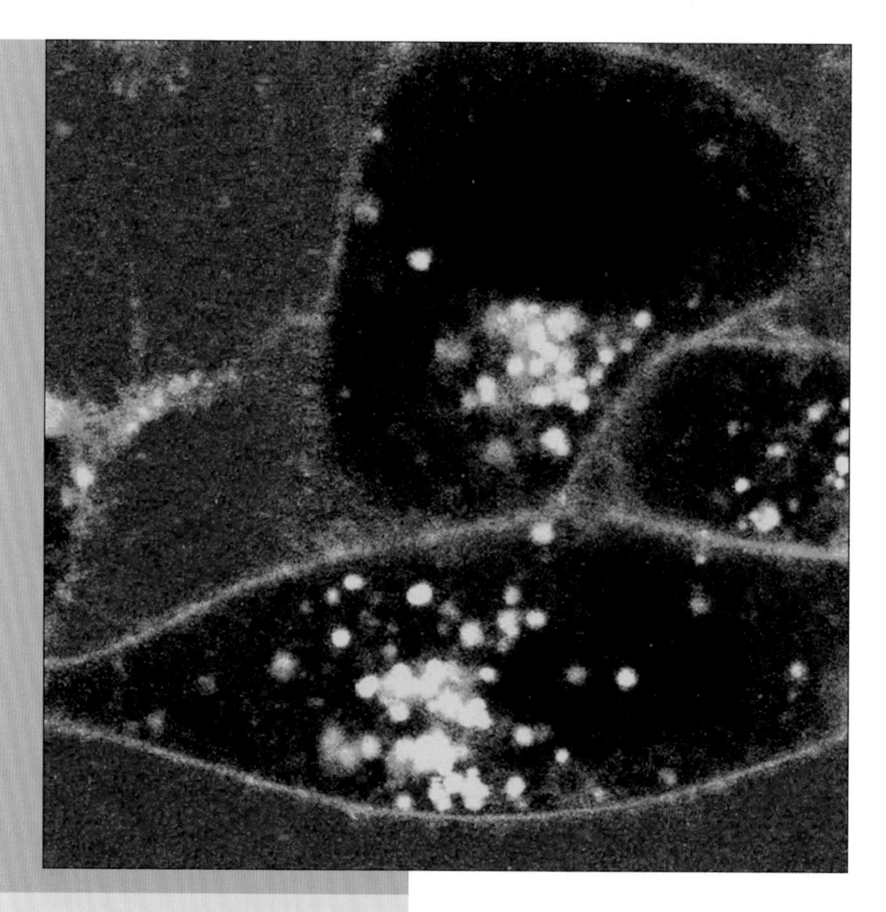

This image was taken using fluorescence resonance energy transfer and reveals how fluorescent probes mimicking the cancer drug paclitaxel enter a cancer cell after being delivered by tiny synthetic spheres called polymer micelles. The study showed that fluorescent probes were released from the core into the tumor cell, and how the micelles break down in the blood before they have a chance to deliver the drug to cancer cells. *(Weldon School of Biomedical Engineering, Purdue University)*

in the protein's fluorescence. This can show whether the substance has bound and whether the protein's activity changes as a result.

If two molecules have been marked with different tags, a technique called fluorescence resonance energy transfer (FRET) shows when and where they bind to each other. (The method is also known as Förster resonance energy transfer, named after Theodor Förster, a German physical chemist who discovered the phenomenon.) Binding brings the fluorescent modules of the two proteins very close to each other. If one of them is excited by a laser and gives off energy, some of it may be absorbed by the partner molecule. This can be seen by reading the signal. In an experiment of this type, Bastiaens and his colleagues used FRET to show that a signal can be sent extremely rapidly into the cell by a receptor called ErbB1, which receives signals from proteins such as the epidermal growth factor (EGF). Because receptors are sometimes triggered by mistake, multiple receptors usually have to be triggered to provoke a strong response from the cell. But a study by Bastiaen's lab, published in *Science* in 2000, showed that EGF can evoke a powerful response even if a single receptor is activated; the stimulated receptor generates a "wave" of signaling that activates other copies of ErbB1 in the membrane.

Until the development of fluorescent proteins, light microscopes had lagged behind biology's entry into the molecular age because they were not powerful enough to resolve single molecules. That has changed with the introduction of methods such as FRAP and FRET. None of the methods capable of attaining higher resolution, such as X-ray crystallography, can observe molecules in their real everyday surroundings. Although researchers have become extremely clever about getting around this problem, it will always be necessary to study the elements of life in the context of living cells.

ELECTRON MICROSCOPES AND CRYO-ELECTRON TOMOGRAPHY

Ideally scientists would like to develop a sort of "GoogleEarth" of the cell, where microscopes can zoom in on locations, pinpointing what molecules are active there and what they are doing. But even the most powerful electron microscopes lack the resolution to see single molecules. They can, however, catch a glimpse of very large proteins or protein machines. That might at least allow scientists to look at an image of a cell and point out the locations of particular machines. Doing so would require a good idea of the shape of the machines, however—to know what to look for—and until recently no one knew what they looked like. Bioinformaticians like Rob Russell of the European Molecular Biology Laboratory in Heidelberg have been working with data from Gavin and her colleagues to try to solve the problem. They take structural information about single proteins in yeast cells and try to fit the molecules together by computer to make an educated guess about the overall shapes of machines. These "puzzles" could be used as a guide to interpret high-resolution microscope images. In the meantime, electron microscopes are being used in new ways to explore the cell. One of these methods, cryo-electron tomography, is now producing detailed maps of the cell interior that will be extremely useful when looking for protein complexes.

Electron microscopes were invented in the 1930s by researchers working for the Siemens company in Germany. The first really useful instrument was developed in 1938 by Eli Franklin Burton (1879–1948), with students James Hillier and Albert Prebus, at the University of Toronto, Canada. Electrons have a much shorter wavelength than photons, which gives the instrument a resolution that is a thousand times better than light microscopes; some electron microscopes can magnify samples up to 2 million times. To use the instrument, samples have to be examined in a vacuum, because even air scatters electrons and would create an image with too much noise. This means that living cells or tissues cannot be studied. Another problem is that exposure to the electron beam gradually destroys the sample. Various methods have been developed to reduce noise and increase the lifetime of the samples, including coating them in gold or other metals and flash-freezing them. Currently several types of instruments are in use, including the transmission electron microscope (TEM) or the scanning electron microscopes (SEM). The TEM works by sending a beam of electrons at a sample and capturing electrons that have passed through. The SEM picks up low-energy electrons that have been excited by the beam.

One way to increase the detail of the images is to eliminate noise, and in the 1990s several laboratories began using computer software to do this. The idea was to take hundreds or thousands of images of the same type of object and superim-

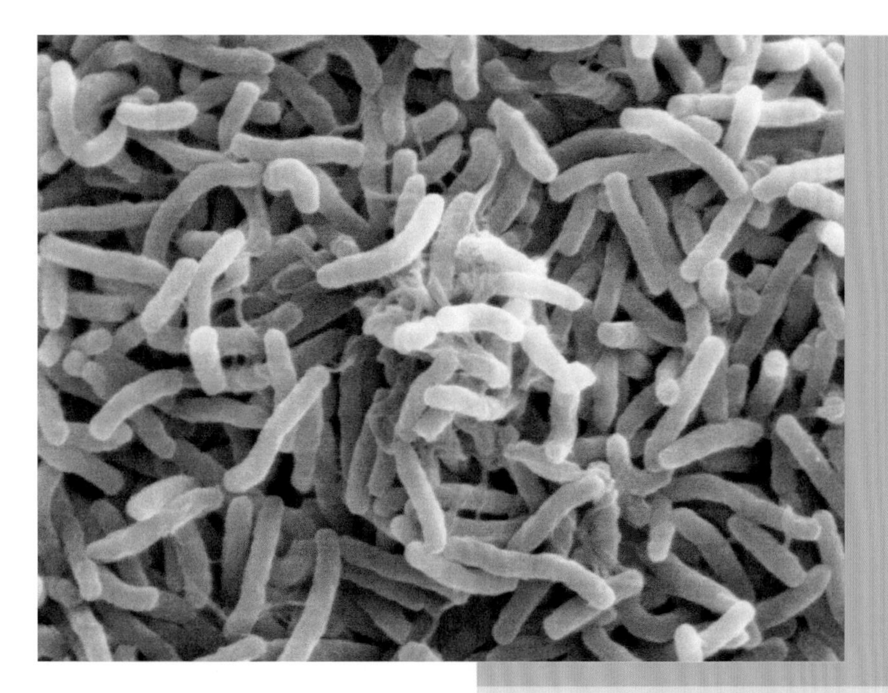

A scanning electron microscope image of *Vibrio cholerae*, the bacteria that cause cholera *(Molecular Microbiology)*

pose them. Since noise was random, the software looked for common elements in the images and eliminated everything else. For the method to work well, scientists needed a huge number of images that not only contained the same object but showed it in precisely the same position. Some of the earliest studies were carried out on the icosahedral viruses. These infectious agents take their name from their shape, a figure with 20 sides, much like a soccer ball. The surfaces of these viruses are very symmetrical, which means that they look virtually the same when turned in many different directions. Since an infected cell often holds thousands of copies of such viruses, a large number of samples could be seen in a single cell. This method has produced very sharp images—at a resolution higher than nine angstroms—of viruses such as the herpes simplex virus-1.

Another method to obtain a cleaner, sharper view of objects within cells—and to see them in three dimensions—is to take multiple pictures of the same object from different angles and combine them. This is the principle behind cryo-electron tomography. Tomography is a method used with different kinds of waves, including X-rays and ultrasound, to focus on a precise depth inside an object, like adjusting the focus on a camera lens so that objects in the center of a room are sharp while things closer or farther away are blurry. Since the waves penetrate the surface, ultrasound tomography can provide doctors with an image of an embryo in the mother's womb, and cryo-electron tomography reveals objects inside cells. A sample is placed in the electron microscope and is turned so that the electron beam passes through it from different directions. The images are then combined into a single, three-dimensional picture.

Stephen Fuller's laboratory at the University of Oxford in Great Britain has been using the method to study how cells build HIV (human immunodeficiency virus). During infection, HIV slips its genetic information into the DNA of a host cell. When that information becomes active, cells begin producing the raw materials needed to make new copies of the virus. The components are produced in various regions of the cell and have to be brought together, assembled, and packed with new copies of the virus's genetic code. This process has to be carried out so precisely that it usually fails; most of the copies of HIV made by an infectious cell are wrongly built and are no longer infectious. But only a few have to succeed to invade other cells, eventually destroying the health of the host and moving on to new victims.

Fuller and his colleagues have been carefully studying the process by which HIV is built, hoping to discover a few key events that make the difference between the construction of infectious versus harmless copies. Their aim is to find a part of the process that can be targeted by new anti-HIV drugs, turning all copies of the virus into the noninfectious type. In

(opposite page) An image of a complete yeast cell made using cryo-electron tomography. Visible are the plasma membrane, microtubules and light vacuoles (which appear as green), the nucleus, dark vacuoles and dark vesicles (which are gold), mitochondria, and other cellular structures. *(Johanna Höög, EMBL Heidelberg)*

2006 Fuller's group used cryo-electron tomography to track the construction of the core of the virus, a cone-shaped, internal compartment that contains HIV RNA and proteins. The cone is made of proteins, and they are assembled inside a sphere of

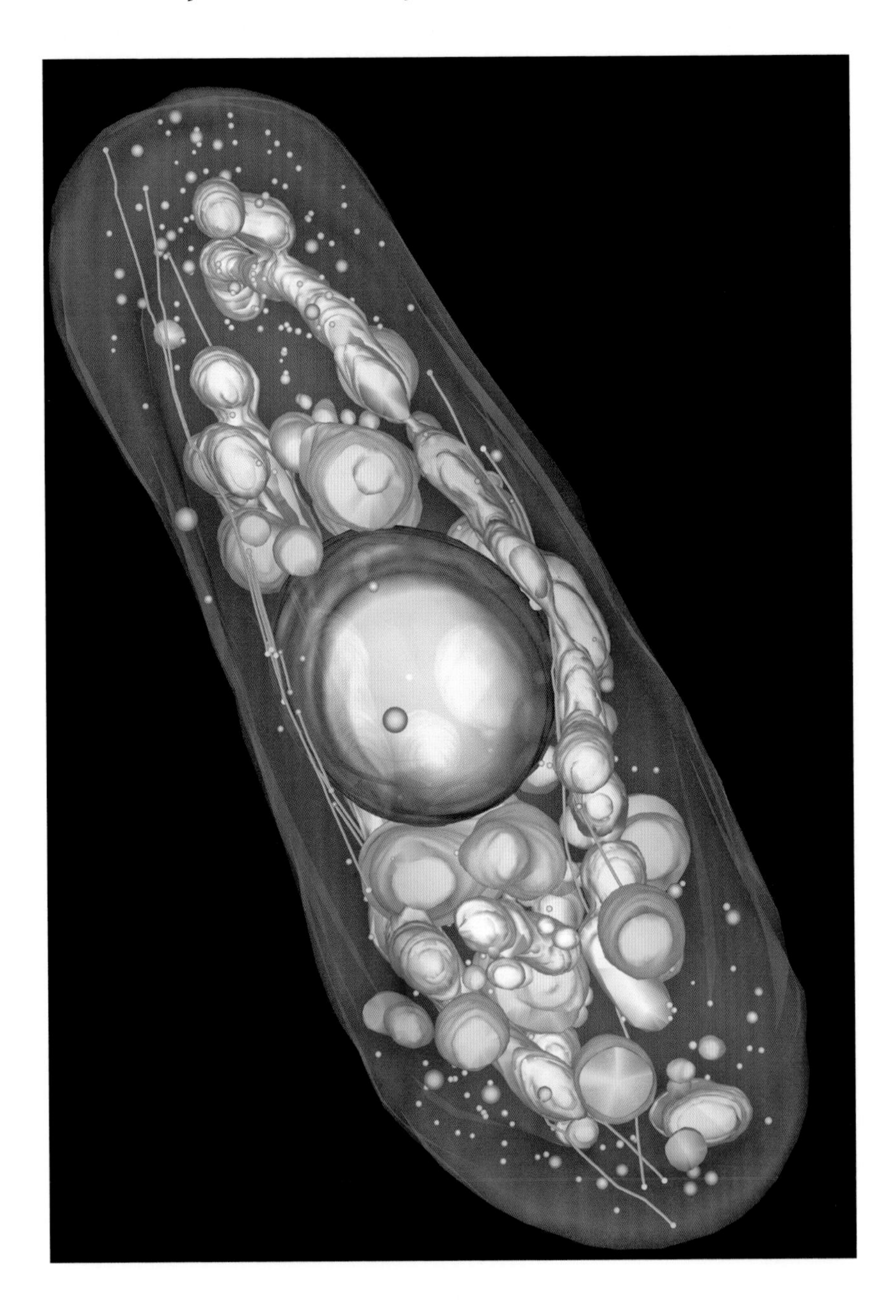

membranes that will become the outer capsule of the virus. In contrast to other aspects of HIV, the cone comes in various sizes and shapes; Fuller was curious about how the cell could tolerate these differences and yet still manage to make a successful, properly shaped, sealed compartment. His lab discovered that the cone begins taking shape at the narrow end and builds outward. Eventually it reaches the membrane, and the ends bend and fuse to seal off the compartment.

Another 2006 study using cryo-electron tomography, carried out by Dirk Schüler's laboratory at the Ludwig Maximillians University in Munich, revealed a strange new structure in "magnetotactic" bacteria. These organisms are widely spread throughout the oceans. Their unique characteristic is the magnetosome, a tiny internal organ that contains crystals of the mineral magnetite. The crystals are magnetic, and they help give the bacteria a sense of direction. The cells use the Earth's magnetic field to navigate toward tiny pockets of oxygen on the oceanfloor. Schüler and his colleagues studied the bacteria using cryo-electron tomography, hoping to discover how the cell controlled the locations of magnetosomes. Although they ought to form clumps because they attract each other magnetically, the organelles are held in a straight line in the cell. The microscope images revealed a new, fiberlike structure made of proteins that held them into place. Another of the cell's proteins appears to glue the magnetosomes to specific positions along the fiber.

Since the "invention" of molecular biology in the 1940s and 1950s, researchers have discovered a great deal about the structure and behavior of individual molecules. The goal of cryo-electron tomography and other types of microscopy is to produce what Wolfgang Baumeister, of the Max Planck Institute for Biochemistry in Martinsried, Germany, calls the "molecular sociology of the cell." In a review published in the journal *Nature* in December 2007, Baumeister asserted that "hybrid" techniques available to today's scientists should allow scientists to discover the internal architecture of protein machines and to study their dynamic interactions in the cell.

2

How the Cell Stores and Uses Information

Every cell that is alive today is the descendant of an organism known as the last universal common ancestor, a cell that may have lived 3 billion years ago. Its descendants evolved into the three great branches of life known today: bacteria, archaea, and eukaryotes. The last group includes all plants and animals as well as a wide range of one-celled organisms such as yeast. Scientists know that all of these organisms share a common ancestor because of common information found in their genomes. Despite the many changes and innovations that have produced at least hundreds of millions of species, each cell is a living representation of the working-out of information it received from its parents, their parents, and its most remote ancestors. The ability to store and protect hereditary information was a precondition for life on Earth. The topic of this chapter is the way organisms access and use that information.

The central storehouse of information is DNA, the cell's archive of hereditary information that is simultaneously its library of recipes for other molecules. *Messenger RNAs* (mRNAs), which are encoded in genes, are temporary information storage devices because they act as templates for the creation of proteins. The set of mRNAs continually changes, as does the population of proteins, but each type of cell has a core set of molecules that it needs to make to attain a certain shape, divide at the right time, and develop properly. This, too, is an important type of cellular memory.

While all life on Earth is based on DNA, RNA, and proteins, there are some important differences in the way very different types of organisms handle these molecules. In this book the focus will be on eukaryotic cells: cells with a nucleus, the type making up humans, other animals, plants, and some unicellular organisms like yeast. The other two major branches of life, bacteria and archaea (also single-celled creatures), generally produce proteins in a similar but simpler process.

THE DOUBLE HELIX

When the young American James Watson arrived in Cambridge, England, in 1951, many scientists were finally becoming convinced that the molecule called DNA contained the genetic code. It had been more than 80 years since a German chemist named Friedrich Miescher (1844–95) had isolated the molecule from pus in the bandages of wounded soldiers. Experiments carried out in 1928 by Frederick Griffith (1879–1941), a medical officer in London, showed that one type of bacteria could be given the characteristics of another strain through a transfer of molecules from one to another. He might have gone on to prove what type of molecules were responsible, but the work was interrupted by his death in the wartime bombing of London. The work was followed up by another scientist who was reaching the end of his career. Oswald Avery (1877–1955) had been working on a vaccine for pneumonia for nearly 35 years at the Rockefeller Institute for Medical Research in New York, a project made unnecessary by the discovery of antibiotics. In 1940 he took up where Griffith had left off and demonstrated that the DNA of bacteria contained the "transforming" information. But scientists had no idea of how DNA could carry information from one cell to another, or whether the same was true for other organisms.

Scientists hoped that understanding the chemical and physical structure of the genetic material might provide answers. In

1937 William Astbury, a physicist specializing in textiles at the University of Leeds, England, turned his attention to DNA. He had been using X-rays to obtain diffraction patterns of wool and other fibers, and now he obtained a set of X-ray images of DNA fibers that clearly showed that the molecule had a regular helical structure, like a spiral staircase. The four chemical building blocks that made up DNA—adenine, cytosine, guanine, and thymine—had to fit together with phosphate and sugar molecules to make this shape. The situation was like having pieces of a puzzle made of blocks of wood that had to be assembled to make the helix. An international race began to put the pieces together.

The problem intrigued Linus Pauling, a young "star" of American science with degrees in both chemistry and physics, who was bringing his knowledge of the new physics of quantum mechanics to bear on chemical and biological problems. His laboratory at the California Institute of Technology began carrying out similar experiments. He had already used Astbury's X-ray methods to make crucial findings about protein structure, discovering alpha helices. This work earned him a Nobel Prize in chemistry in 1954. Eight years later he became only the second person in history (alongside Marie Curie) to win another Nobel in a second category. This time it was the 1962 Nobel Peace Prize, for his efforts to stop the testing of nuclear weapons aboveground. He had become interested in political issues through the efforts of his wife, Ava, a peace activist and human rights advocate. His activism came at a price: In 1952 he was denounced as a communist before Senator Joseph McCarthy's committee in the U.S. Congress. That led the government to refuse him a visa to attend a scientific meeting of the Royal Society in London later in the year. One of his colleagues, Robert Corey, went instead. During the trip Corey met with a young researcher named Rosalind Franklin, who was using crystallography to investigate DNA.

It is hard to guess what might have happened had Pauling attended the meeting, but he might have obtained data that would have helped him create an accurate model of DNA. A year earlier he had published a paper that incorrectly proposed that DNA was built of three twisting strands.

Watson and his colleague Francis Crick were also working on DNA in Cambridge. They had recently proposed a hypothetical structure of the molecule that was so wrong it was embarrassing. Their boss, Sir Lawrence Bragg ordered them to stop the project. Working with his father, William Henry Bragg, Lawrence had done groundbreaking studies using X-rays to interpret the first structures of minerals such as iron pyrites and sodium nitrate. The two shared the 1915 Nobel Prize in physics, making Lawrence, at 25, the youngest person ever to win the award (his record has held ever since). Watson was also a young prodigy, having entered the university at the age of 15, but Bragg gave the two men little chance of success. Neither was an expert in chemistry. How could they compete with researchers like Pauling, one of the most famous scientists in the world, who was also working on the molecule? But when Pauling's structure appeared in print it contained many of the mistakes that Watson and Crick had made. Bragg allowed them to start up again. But by this time the landscape had changed. Franklin, a young British crystallographer, had discovered an error in the way Astbury and others had been collecting X-ray data about DNA. She joined the laboratory of Maurice Wilkins, a physicist in London who had begun to work on DNA. Her experiments gave a much clearer image of the molecule that would become essential in solving its structure.

Franklin had figured out that DNA came in two forms: "A" and "B," a "dry" and a "wet" form, respectively. Under humid conditions more hydrogen atoms docked onto the molecule, and it changed its shape. She realized that in the images taken by Astbury and Pauling, the two forms were mixed. Their images were superimposed and blurred. Using only the B form, Franklin obtained the sharpest-ever images of DNA. She began trying to interpret what they said about its structure but interrupted the work to go on vacation. While she was gone, Wilkins, without her permission, showed some of her X-ray images to Watson and Crick. One look at a photograph of the B form was enough to convince them that DNA formed a double helix, rather than the triple-stranded structure they had proposed earlier.

Sugars, phosphates, and bases had to be fit into this shape. Franklin's X-ray data showed the circumference of the helix and how high the steps were. The latter value—the space between neighboring nucleotides—was 3.4 angstroms, or 3.4 times the diameter of a single hydrogen atom. The width of the entire staircase was 27 angstroms. She also discovered the space needed for the helix to make a complete turn: 34 angstroms, so it took 10 nucleotides to make one trip around the helix.

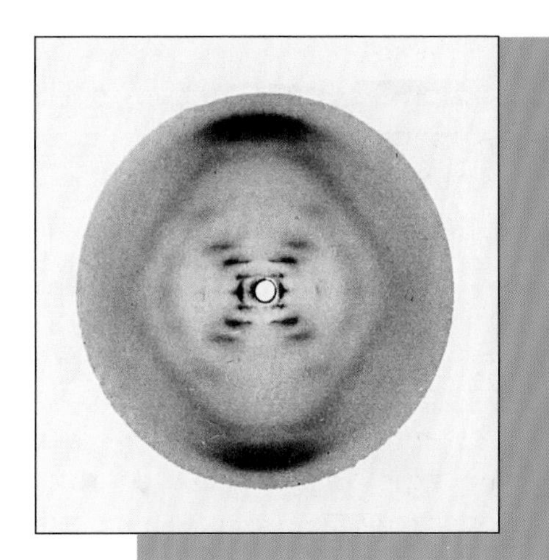

An X-ray diffraction pattern of the "B" form of DNA, using sodium deoxyribose nucleate from calf thymus, taken by Rosalind Franklin and R. G. Gosling. *(AIP Emilio Segrè Visual Archive)*

But this did not show which piece went where. Watson made cardboard cutouts in the shape of the four bases and began working on the puzzle. No matter how he tried to attach them to each other, something always failed to fit. One problem was the angle at which the stairs turned; the rotation between two steps had to be 36 degrees. Neither Watson nor Crick could get the pieces to match Franklin's data.

Then Watson's office mate—ironically, a former student of Pauling's—told him that bases existed in two different chemical forms, with slightly different shapes. Following the chemistry textbooks, Watson had been using a form with an extra oxygen atom. Now he remade his cardboard shapes based on the type without the oxygen. He was idly fitting them together when he had a sudden revelation: when A snapped onto T, it had almost exactly the same size as G fit to C. Fit together, their size matched the dimensions shown by Franklin's X-ray photographs.

It also explained a discovery that had been made by Erwin Chargaff (1905–2002) in 1950. Chargaff was an Austrian

James Watson (left) and Francis Crick, with their physical model of the DNA double helix *(Barrington Brown/Science Photo Library)*

biochemist who had immigrated to the United States and ended up at Columbia University. Avery's work on pneumonia convinced him that DNA contained the genetic code. If so, each species ought to have its own particular recipe of bases, and Chargaff began trying to detect it. He found, for example, humans had a much higher proportion of the base G than flies did. Along the way he discovered a strange phenomenon: Within any single species, the amount of A was almost identical to that of T, and amounts of G were almost identical to that of C. This made perfect sense in a model in which four bases only formed two pairs. Chargaff did not make the connection; Watson, toying with cardboard models, suddenly saw its meaning.

Complementary bases join to make the "steps" of the double helix staircase, and they snap onto each other using hydrogen bonds. As explained in chapter 1, hydrogen is a very simple atom that easily links to a neighboring atom that lacks an electron. The surface of a nucleotide has a pattern of charges that perfectly fits the pattern of its complementary base, the way that a puzzle piece snaps onto its neighbor. Watson and Crick quickly wrote up their findings in a paper called "A Structure for Deoxyribose Nucleic Acid" and submitted it to the journal *Nature.* It was published in record time, just three weeks later. Key X-ray data from Franklin and Wilkins were published in a separate article in the same issue, which appeared on April 25, 1953. The new model convinced everyone. The structure of DNA had been solved, and with it, the mystery of what genes were made of. Its importance was so widely recognized that just 10 years later Crick, Watson, and Wilkins were awarded the Nobel Prize in physiology or medicine.

FROM DNA'S STRUCTURE TO ITS FUNCTIONS

The double helix is a perfect example of how a molecule's form determines its functions. In most cases, when a cell divides, whether it is a bacterium or part of a human body, each of its offspring needs a complete copy of the organism's hereditary information. That information must be copied accurately, but the system is not perfect—otherwise there would have been no evolution. The double helix revealed the form in which hereditary information was stored (DNA). It also showed how that information could be copied: If the two strands were pulled apart, each strand contained enough information to rebuild its partner, because each of the bases could only bind to one partner. The model offered fulfilled the requirements of Franklin's X-ray images and Chargaff's data. It even explained how mutations could arise. As Watson and Crick pointed out in their landmark

Working with Rosalind Franklin: A Personal Account

In 1963, shortly after receiving the Nobel Prize in physiology or medicine, James Watson published his version of the discovery of the structure of DNA in a book called *The Double Helix*. The book revealed for the first time how Watson and Crick had been shown Rosalind Franklin's X-ray data and how important this had been to their solution. In retrospect historians have wondered whether Franklin—rather than her colleague Maurice Wilkins—should have shared the Nobel Prize with Watson and Crick. The question is purely academic because the prize is only awarded to living scientists, and Franklin had died in 1958. But it has called attention to the fact that great accomplishments in science are almost always made possible by the work of many people in the background who rarely receive the notice they are due. It has also renewed interest in Franklin's life and career. Ken Holmes, introduced in chapter 1 as a pioneer in the use of synchrotron radiation,

Rosalind Franklin was an avid hiker who frequently spent her vacations hiking in Europe. *(From the collection of Jenifer Glynn, National Library of Medicine, National Institutes of Health)*

did his doctoral degree with Franklin in the late 1950s. The following shares some of his personal memories of that period.

> I did my first degree in Cambridge from 1952 to 1955, during the period that Watson and Crick were solving DNA and Max Perutz and John Kendrew were solving the first crystal structures of the proteins myoglobin and hemoglobin. However, I did not have any contact with that work at the time. When I finished my degree, I wrote around to find someone willing to take on a young, would-be crystallographer to do a doctoral degree. J. D. Bernal, who was professor of physics and head of the department at Birkbeck College London, passed my letter on to Rosalind Franklin, who had just received a grant from the British Agriculture Research Council (ARC) to work on the structure of the tobacco mosaic virus, or TMV, a simple rod-like RNA virus that infects tobacco and related plants (and was actually the first virus ever discovered). She offered me a research assistantship that I gladly accepted. Later, her three-year grant from the ARC was not renewed on the progressive grounds that the ARC didn't really fund women! Fortunately for me (since it paid my salary) she then became one of very few foreign people to receive a grant from the National Institutes of Health in the United States.
>
> Bernal set her up in a rather decrepit Georgian terrace house in Torrington Square, Bloomsbury, attached to Birkbeck College. These houses have a basement where presumably the servants had

(continues)

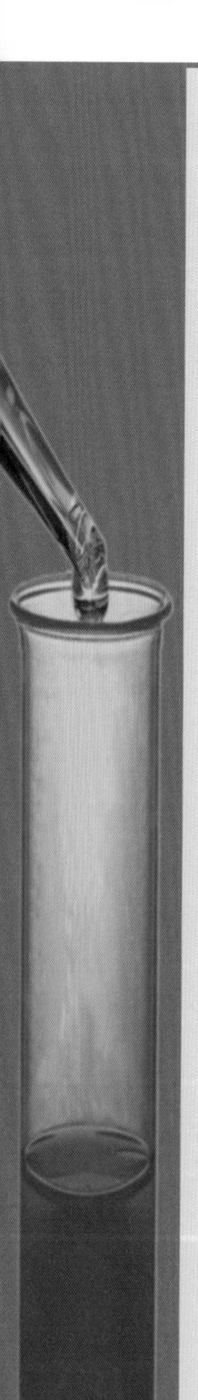

(continued)

worked (now used for X-ray apparatus and the workshop), and garret rooms four stories away at the top where they might sleep. Rosalind had been given one of these garret rooms (that also leaked whenever it rained). Naturally, her apparatus was mostly in the basement. Aaron Klug came as a British Empire Fellow initially to work with Harry Carlisle in Torrington Square on the structure of a protein called ribonuclease. Aaron was of theoretical bent, and he managed to show that Harry Carlisle's proposed structure of ribonuclease was wrong, which led to his banishment from the main floor to a garret room next to Rosalind. This turned out to be a happy circumstance since they inevitably met on the stairs and subsequently established a very fruitful collaboration on virus structures. There ensued a flood of papers from Franklin and Klug. I joined Rosalind in 1955 as a research student (I didn't get put in the garret) to work directly with Rosalind on the structure of TMV. Bernal had shown that solutions of the long rodlike TMV particles spontaneously form orientated liquid crystals in response to a magnetic field [liquid crystals are nowadays of great importance as the basis of LCD displays, although they don't normally use TMV!]. Such liquid crystals give detailed X-ray diffraction diagrams known as fiber diagrams. Rosalind's plan was to obtain the atomic structure of TMV from such X-ray fiber diagrams. That took a long time. Years later, in Heidelberg, my lab eventually got it down to four angstroms resolution, and colleagues of mine went on to finish it at 2.5 angstroms and were able to build an atomic structure of a virus (one of the first virus structures known). What makes this work unique is that it was not done from protein crystals, the way most structures are solved, but from X-ray fiber diagrams, just the type of X-ray diffraction that had been used to figure out the structure of DNA. In fact, DNA and TMV were for a long time the only biological structures that had been solved by X-ray fiber diffraction. The work necessitated a host of technological advances, both experimental (largely originating from Rosalind) and theoretical (from Aaron) and the kind of detailing that originates from the guy at the coal face. After this ordeal I continued to work on fibers, mostly fibrous proteins called actin and myosin that play an important role in muscle contraction. Actin does not form crystals, so the only way to work with it was as a liquid crystal. My training on TMV stood me in good stead.

In all I worked with Rosalind for three years. She was not easy to talk to; she suffered fools not at all. Moreover, she had a difficult time in casual conversation. I think there is a certain type of English person who is somewhat inhibited by being in England, who is bothered by all of the sociological rules and regulations that come along with it. Conversational "gambits" are an important part of that, for example, being able to manage a conversation with strangers at table, and in such circumstances Rosalind did not perform well. At one stage in the common room for lunch she was reported to have interjected, "I hear it is a good year for mushrooms." In France, where she had worked for four years, these things are much easier, and in a French ambience Rosalind was much

(continues)

(continued)

more relaxed. Particularly in French she could be animated and light-hearted. Shortly before her death she came to dinner with us. She charmed us with sweet, sad stories from her four years in Paris just after the war, when obtaining food was often problematic. One stays in my mind. Rosalind was rather proud of having obtained a bottle of milk (then uncommon in Paris) that she duly stored on the window ledge of her (garret) room. It was winter, the milk froze, and sadly the glass broke. The milk ran into the gutters!

She never mentioned her time at Kings College. Jim Watson's picture of Rosalind is a lamentable misrepresentation. What particularly irritated me was Jim's total misapprehension of Rosalind as a woman. She was no bluestocking but a smart professional woman who happened to be a scientist. In the face of considerable difficulties she quickly built up a small, loyal, and dedicated group. Moreover, perhaps inevitably, in the last year of her short life her bravery and optimism led to a deep bonding.

She initially had mixed feelings about my wife, Mary, whom I was courting at the time, because she thought that Mary might be taking too much of my time, which would not be good. Nevertheless, Rosalind gave us a nice wedding present. Furthermore, we worked well together, and she taught me lots. Perhaps we both had the same kind of attitude; we were both rather perfectionist. Our collaboration led to an important paper on the symmetry and structure of TMV that we published in *Acta Crystallographica* in 1958. Apparently, when the paper was submitted, it was reviewed rather critically by Francis Crick, who did not believe that anyone could have obtained such accurate measurements. Indeed, what we were doing was very difficult; to start with you need a very well orientated sample of TMV. To get orientated liquid crystals of TMV you have to suck a concentrated solution of the virus into a capillary tube, where it starts forming an oriented gel. When this has been done you have to seal the capillary tube with a hot flame without blowing all the liquid out again. Rosalind was very good at these finicky technical sorts of things. To get the diffraction data we used evacuated X-ray cameras, special X-ray sources, and bent quarz monochromators, all designed to keep out scattering from extraneous sources (such as air) so that we could record the data accurately. All this technology was on the edge of what was then practicable and therefore neither easy nor reliable. Nevertheless, we got it right. Much of what I learnt from Rosalind about X-ray optics later became incorporated in the design of the first X-ray beam lines for synchrotron radiation [see sidebar, chapter 1].

Rosalind died in 1958 of cancer, which was a dreadful loss, the loss of a fascinating person and a great scientist who surely would have gone on to do even more important work. The group was taken over by Aaron Klug, who was a superb theoretician and went on to win the 1982 Nobel Prize in chemistry for his work on crystallography and electron microscopy. I finished up my Ph.D. in 1959 and went off to do a postdoctoral fellowship at the Children's Cancer Research Foundation in Boston, where I began working with muscle. I came back to

(continues)

(continued)

Britain to rejoin Aaron Klug's group in the new Laboratory of Molecular Biology in Cambridge, which was duly opened by the queen in March 1962. That autumn the Nobel Prizes were announced: first for Max Perutz and John Kendrew for protein structures, then for Francis Crick with DNA. There was a lot of champagne and partying. The atmosphere was, "Well, Nobel parties happen every week, don't they!" The Nobel laureates bought so much champagne that they began getting literature from wholesalers. It was an amazing lab (to date 13 Noble Prizes), and an amazing time.

paper, in rare cases hydrogen atoms might bind differently to a base, slightly changing its shape. As one strand was copied, it might then attach to the wrong base.

What the model did not show was how the cell used information in DNA to create proteins. Watson and Crick had been discussing a hypothesis that Crick announced publicly, in 1959. Crick called it the "central dogma of molecular biology"—"DNA makes RNA makes proteins"—introduced in chapter 1. The main focus of biology after the discovery of the double helix quickly became working out how the cell carried out these transformations. By about 1970, researchers had worked out the DNA code. Many of the details have only become clear since the completion of the human genome, and others remain unclear, but the following list contains the most important steps as scientists understand them today:

- The DNA in cells is enmeshed in a dense collection of other molecules, mostly proteins; the whole complex is called "chromatin."
- The human genome contains about 3 billion base pairs.

- Only about 1.5 percent of the human genome consists of genes, which contain the instructions for creating proteins.
- Other DNA sequences control whether genes are used to make RNAs. They provide docking sites for other molecules that may repress or activate a gene. Usually this happens because proteins recruit the machinery that transcribes DNA into RNA and escort it to the gene. When they get there, the proteins may help the machine dock, or they may find the site blocked by repressor molecules.
- At any one time, most of the genes in a particular type of cell are silent; they are not being used to produce RNAs. Patterns of active and inactive genes give cells unique identities by changing their chemistry, giving them different shapes, making them receptive to different signals from the environment, and so on.
- Genes are first transcribed into mRNAs in the cell nucleus. RNAs are carried into the outer compartment of the cell, the cytoplasm, to be translated into proteins
- Nearly all genes in humans, other animals, and plants include regions of protein-coding information (*exons*) interrupted by huge stretches of "noncoding" information (*introns*). The average human gene has 8.4 introns; this number varies considerably from species to species.
- Introns have to be removed from RNAs in the nucleus in a process called "splicing." Sometimes specific exons are also removed. This process, *alternative splicing,* allows the cell to create different versions of a protein using the information from a single gene by removing different sets of exons.
- It takes three letters of a DNA (and RNA) sequence—three consecutive bases—to spell one letter of the protein alphabet (amino acids). The three-letter groups are called *codons.* There are 64 possible codons and only 20 amino acids, which means that each amino acid can be spelled by a few different codons. Some of the 64 possible codons do not encode an amino acid. Instead, they act as a signal that shows the cell where the protein-encoding part of an RNA ends and are called *stop codons.*

Every year scientists discover new ways that cells regulate the pathway between genes and the proteins they encode. Some of the latest discoveries involve how the cell creates mRNAs but interrupts their translation—discussed later in this chapter. To understand what goes on in cells it is essential to examine what happens at each step of this process, and it all begins with the gene itself.

THE STRUCTURE OF SINGLE GENES

The accompanying image is adapted from a database of the human genome created and maintained at the University of California, Santa Cruz (UCSC). The image represents BRCA2, a gene that has been linked to breast cancer. The normal function of this gene is to help repair chromosomes that have been damaged by radiation or other environmental factors. It is called a "tumor suppressor gene" because if it is missing or damaged, humans suffer a greater risk of developing cancer. This diagram of BRCA2 will be used as an example to show how complex genes are.

The gene is found on chromosome 13, which can be seen as the striped band at the top of the image. The stripes really exist; they can be seen when chromosomes have been stained in a certain way. They always appear in the same places and act as landmarks, like highway mileage markers. Since the staining technique was invented in the early 20th century, the stripes have been used to map the locations of specific genes.

The small red bar shows the location of BRCA2 on the chromosome. At this resolution no details can be seen, so the region has been greatly magnified below. In the close-up view, the complete gene (including its introns) appears as a thin horizontal line. This is still a very basic view of the structure; it does not reveal, for example, the DNA sequence of the gene. (Seeing that requires zooming in about 1,000 more times.)

The image reveals the most important features of the gene. The heading "size" at the top says that the sequence containing the gene spans 84,193 base pairs. Only a fraction of this infor-

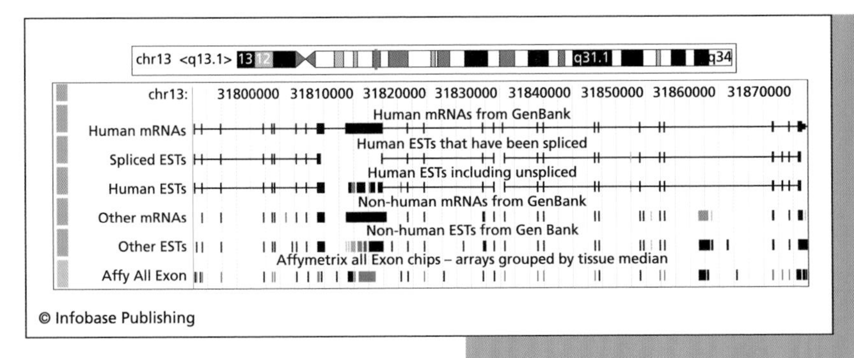

This image, adapted from the University of California, Santa Cruz's Internet genome database, reveals the complex structure of the BRCA2 gene, which has been linked to breast cancer. The text contains a detailed explanation of how to read the chart.

mation, however, is actually used to make BRCA2 protein. This can be seen from the diagram of the gene itself: the thin black horizontal line interrupted by vertical hash marks. Each hash mark represents the position and size of a protein-encoding exon. BRCA2 has 25 of these regions; everything else is noncoding material that gets removed through splicing. On the far right the gene ends in a small symbol shaped like a hat turned on its side: This represents the last exon, followed by an "untranslated region" (discussed later in the chapter in the section on RNA).

The arrows (>) along the line show the direction that code is read to produce RNA. The double helix has two strands (known as "Watson" and "Crick"). Some genes lie on one strand, and some, on the other. More and more cases have been discovered where the same region of both strands is used to produce different RNAs, running in opposite directions. So far there is no evidence that this happens with BRCA2, but if that turns out to be the case, the database at UCSC will be updated, and this information will appear in the entry for the gene. Usually the strand opposite a gene does not encode an mRNA. Cells produce a huge number of small RNAs with other functions; they are discussed later in the section on RNA.

The image also shows expressed sequence tags, or ESTs, that have been found in association with BRCA2. An EST is a short RNA molecule that has been discovered in an experiment, proving that the gene is actually transcribed into RNA. ESTs for the same gene may contain different sets of exons, which show that the molecule can be spliced in different ways. By comparing the sequence of the RNA with the gene, scientists can see whether regions of the sequence have been skipped, which points out the locations of introns and exons. ESTs can also reveal whether mutant forms of a protein exist.

Some regions of the chromosome are labeled as "putative" genes because they appear to contain all the instructions needed to make an RNA and protein, but these molecules have not yet been found in cells. Other sequences were genes long ago but have been deactivated by mutations that happened over the course of evolution.

To make this clearer, suppose that the genome is like a huge dictionary of any word that has ever been used in the English language. ESTs are like small fragments of texts cut from specific books or newspapers, collected in order to discover how the people of one time and place really used the language. Some of the words in the dictionary might not be used at all anymore, or they are only used in unusual situations that were not sampled. So they might not turn up even in a huge collection of text fragments. ESTs also reveal varieties of genes within a population, the way that a collection of text fragments might contain American and British spellings of the same word (*center* vs. *centre*). The collection may also contain misspellings, which are like mutations.

One way to discover whether a molecule is involved in a disease is to take samples from patients and search for unusual ESTs. They may reveal that a particular gene is misspelled, that an RNA is missing some of its parts, or that it has been spliced in the wrong way. These variants are also stored in EST databases because they may be helpful in diagnosing and treating new patients. ESTs also reveal the diversity of the human species; complex genes can often be found in hundreds of different forms in a population. Surveys of these variants have already been useful in understanding human history and evolution.

As complex as this diagram of BRCA2 is, it does not reveal everything about the gene; for example, it does not show the location of other DNA sequences that control when and where BRCA2 becomes active. Single genes are often associated with many such regions, which may be located far from the gene itself, and many remain to be found. How they affect the gene is discussed in the next section.

ACTIVATING A GENE

A cell's identity and behavior are governed by patterns of active and silent genes. Producing RNA from a gene is a complicated procedure that usually involves rearrangements of the DNA strand and the participation of a dozen or more molecules. In human cells an enzyme called "RNA polymerase II" is responsible for reading DNA sequences and using the information to assemble a new RNA molecule. But it does not do this job alone. The steps in the procedure are usually the following:

- A transcription factor (gene-activating protein) called TFIID binds to a sequence near the beginning of the gene. This area is called the TATA box because it contains a high percentage of T-A base pairs.
- When TFIID binds, there are changes to its own shape and the form of the DNA strand, making it possible for other transcription factors to bind.
- When the right factors have been assembled, their combined shape forms a docking site for the RNA polymerase II.
- The characteristics of the site and the molecules bound to it tell the polymerase where to begin making an RNA and where to stop.
- The polymerase can only read and work on one strand, so the two strands of the double helix have to be pulled apart. This begins at the TATA box.
- The polymerase moves around the exposed strand in a spinning motion, a bit like the way a wing nut spins as it is moved along a screw. As each base is read, the polymerase finds a free complementary nucleotide and

attaches it at the end of the growing RNA molecule. Behind it, the double helix closes again.

- A termination sequence tells the polymerase where the gene ends. The RNA falls off the machine or is pulled away by other molecules.

This is the standard procedure for *transcription,* but the details vary from gene to gene. TFIID binds to various types of sequences in slightly different ways, giving access to different sets of transcription factors and changing the conditions under which the polymerase can bind. In front of the TATA box of most genes is another region called a regulator. This is a binding site for yet other proteins that can affect the formation of the transcription complex. And the activation of a gene is usually affected by more distant DNA sequences. Some of these regions promote activation and are called "enhancers." Others hinder their use and are known as "silencers."

Enhancers and silencers, which often lie thousands of base pairs or more from a gene, are docking stations for other proteins. Their influence is usually felt because as the transcription machinery is assembled, the sites are drawn close to the gene in a loop. A protein on an enhancer site may come in contact with TFIID or another one of the transcription factors and create the right shape for a docking site for the polymerase. Silencers have the opposite effect by blocking the activity of the machine.

Until the 1990s it was difficult or impossible to get a direct look at the components of the transcription machine or other protein complexes. This began to change through the development of new instruments and methods, particularly mass spectrometry. If researchers manage to extract an intact machine from the

(opposite page) The text describes in detail the process of activating a gene; this image shows some of the steps. TFIID and several other factors bind to a TATA box motif in the DNA strand, creating a docking site for the RNA polymerase II. When all of the parts of the machine have been assembled, the DNA helix is partially unwound, and RNA polymerase II travels along one strand, making an RNA molecule. The process can be enhanced or inhibited by molecules called enhancers or silencers, respectively, which are bound to other locations in the DNA strand.

cell, they can use mass spectrometry to analyze its components. The method works by cutting the machine's proteins into small fragments. These are "weighed" as they are shot past a magnet

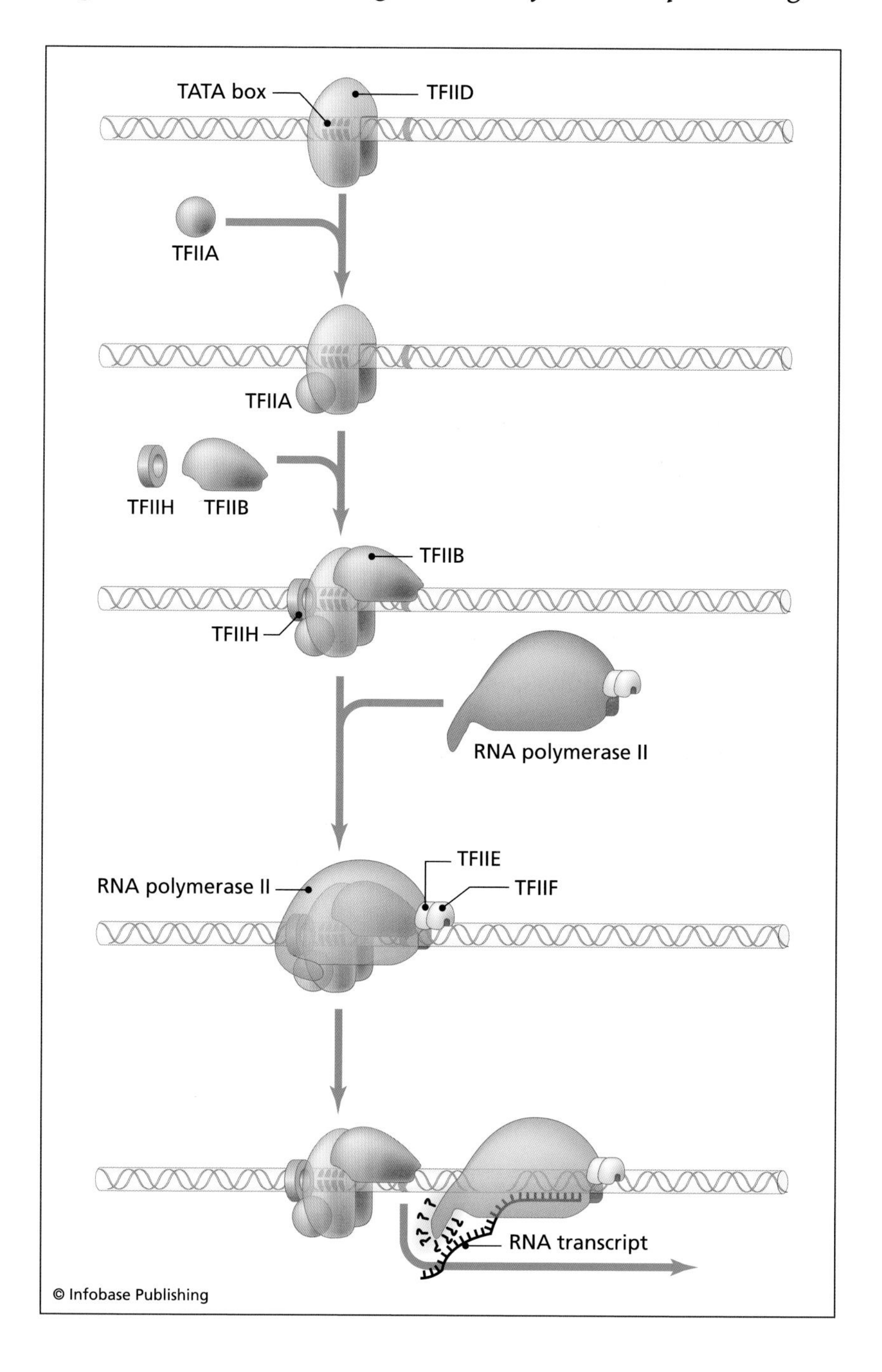

and caught, as they fall, in a detector. Because the fragments carry a charge, the magnet bends them in different ways depending on which amino acids they are made of. The size and charge of a fragment determines where it falls. By analyzing this position, a computer program can figure out the amino acid recipe. Since every protein is made of a unique collection of fragments, the computer can then deduce what protein they belong to.

One use has been to capture transcription activation complexes to discover what they are made of. Breaking down the machines and studying the functions of their parts have given researchers a new view of the subtle ways in which cells control gene functions.

Each step in the activation process is used to control when and where certain RNAs are made. Activating a particular gene may require making new transcription factors or enhancers (whose genes also have to be activated). Once made, a transcription factor has to get access to the gene. That does not happen automatically; sometimes cells silence a gene by locking a transcription factor out of the nucleus. If it does manage to get inside, the protein has to bind to the right DNA sequences, often in just the right combination with other molecules. Even if all of these conditions are met, the process can be blocked by a molecule on a silencer. To be removed, this obstacle may have to be pulled away by yet another protein (a molecule that also has to be produced—involving all the same steps).

The activation of a gene is therefore often a "chicken-and-egg" process that seems to depend on everything else that is going on in the cell. Yet, sometimes the process is much simpler, particularly in the case of genes that are needed often, by most types of cells. Then it may completely depend on the presence or absence of a single transcription factor.

EPIGENETICS: CONTROLLING GENES AT THE LEVEL OF CHROMATIN

Cell division is a major, hurricane-like event that breaks down major structures such as the nucleus and scatters molecules through the cell. Or it is like a massive reset button that would be expected to wipe the slate clean and return the cell to a generic state. In spite of the chaos, cells usually retain their identities and "remember" what their genes were doing before the storm. Over the past decade researchers have begun to find an explanation in the chemistry of chromatin, the mix of DNA and other molecules found in the cell nucleus.

Just before cell division, DNA condenses into tight packages—the chromosomes—which can easily be seen under the microscope. In its normal state the double helix is sprawled through the nucleus in a loose thread, like a ball of yarn that has become unwound. Then the DNA can barely be seen, even by the most powerful electron microscopes. It appears as a very thin strand interrupted by beadlike structures called "nucleosomes." Recent research has shown that these structures play an important role both in packing DNA and controlling gene activity. They also encode a sort of "memory" that keeps cells on track as they divide and differentiate.

This information is called epigenetic ("above" or in "addition to" the gene) because it is not directly encoded in the genome. All cells in the body (with a few unusual exceptions) contain the same genes, but they develop in different ways because of the epigenetic information that a cell passes to its offspring when it divides. The very first cells in an embryo, born from the initial few divisions of a fertilized egg, are identical embyronic stem cells, but very soon most commit themselves to becoming specific types. The 32 cells do not yet look like neurons, muscle, or blood, but they have already activated genes that will eventually push them toward a particular fate. When they divide, their offspring somehow are told to continue producing the proteins needed to guide them on their way.

In 1974 biochemists Ada and Donald Olins of the University of Tennessee discovered nucleosomes and correctly hypothesized that they are built of eight proteins, histones, that bind together in the shape of a small spool. Further studies showed that the DNA thread wraps around each nucleosome 1.65 times. A ninth histone attaches itself to the outside, where it seems to clasp the DNA to the spool. Nucleosomes can be moved, the way that pulling on a thread can turn a spool. They also help with the packing and unpacking of DNA; at the approach of cell division the spools can be packed into tight rows, like packing a cabinet full of cans.

Nucleosomes have an important effect on the activation of genes. It takes 146 base pairs of the DNA strand to complete 1.65 turns around each nucleosome. Bases that are tied up this way are not as accessible to transcription factors or the polymerase needed to make RNAs. The exact locations of nucleosomes on the strand therefore help determine which genes are active and which are silent. If a gene needs to be activated, a nucleosome may have to be moved out of the way. Often this is accomplished by a "chromatin remodeling machine," a group of proteins that dock onto the structure and slide it to a new position on the DNA.

In the 1960s Alfred Mirsky (1900–1974), a researcher at Rockefeller University who had worked on proteins with Pauling, did a number of experiments that showed that chemical changes in histones influence whether or not genes are transcribed. Since then the group of biochemist Michael Grunstein at the University of California, Los Angeles and laboratories around the world have been working hard to understand how these changes happen and what effects they have on gene activity.

Some of the histones have long tails that hang outside the nucleosome, where they can come into contact with other proteins. The particular recipe of amino acids in the tails makes them chemically very reactive. The most common types of changes come when other proteins add or remove small chemical compounds to specific regions of the tails. Adding a compound called an "acetyl group" to the tail of histone 3 usually allows a gene to be activated, while adding a "methyl group" to the same tail normally silences the gene. But in some situations these tags have the opposite effect, and a single histone tail often carries multiple tags. The behavior of a gene probably requires the cell to make sense of several histone modifications spread across several nucleosomes.

Sometimes the cell needs to activate or silence large groups of genes at once, and tagging the histones in whole regions of

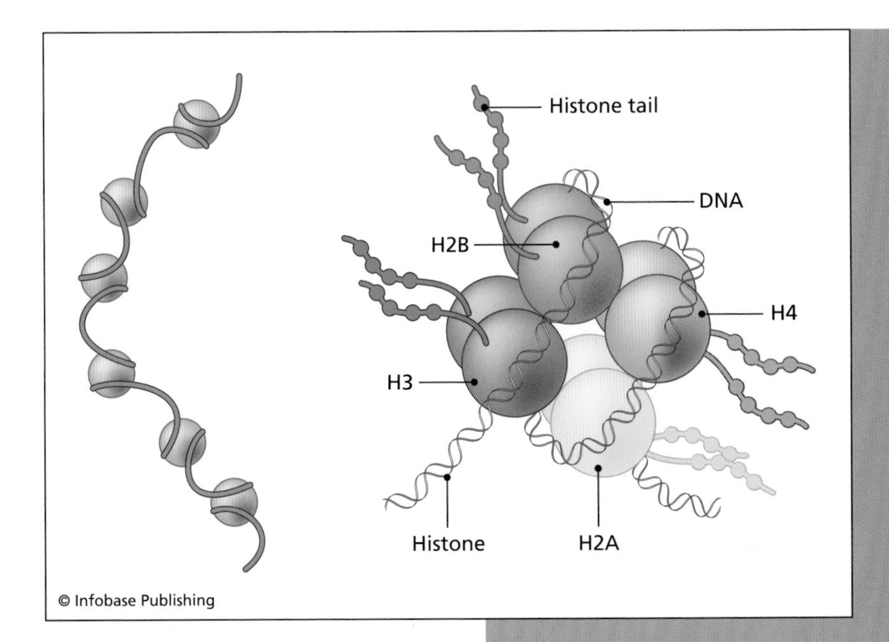

The DNA in a cell's nucleus is wrapped at regular intervals around spool-like structures called nucleosomes (left). Each nucleosome is made of eight core histone proteins (right), whose tails come in contact with other molecules and have an important influence on the activity of genes. A ninth histone acts as a "clasp" that holds the structure together.

DNA with methyl or acetyl groups seems to be one method that it uses. This is important in balancing differences between cells of the two sexes, for example. Females have two copies of the X chromosome, which would normally give their cells double the amount of X chromosome proteins compared to men, who have only one. But in most cases the cells of males and females need the same amount of these proteins; the solution that has evolved in humans and mammals is to silence one of the X chromosomes. Methylation is one mechanism. Others seem to include pulling DNA into regions of the nucleus that serve as "silent zones." In yeast, for example, areas just inside the membrane of the nucleus often act this way. Cells can anchor DNA there by binding it to proteins in the membrane.

How does this create "memory"? When cells divide, not all of the nucleosomes and other proteins attached to DNA are

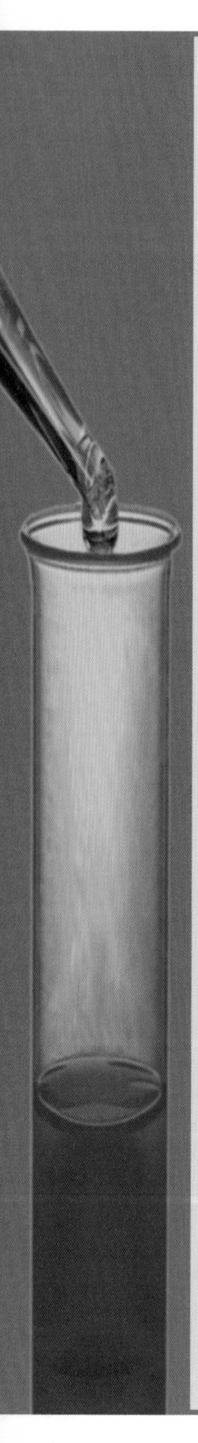

DNA Computers

In 2006 the MAYA-II computer, built by scientists at Columbia University in New York and the University of New Mexico, matched wits against human opponents and scored a win or a draw every time in a game of tick-tack-toe. This might not seem like a huge accomplishment compared to the feats of other computers, such as the 1997 victory of Deep Blue, a chess-playing machine designed by IBM, over world champion Gary Kasparov. But MAYA-II was not an ordinary computer. Instead of circuits printed on silicon boards, the new machine consisted of a plastic tray with small, liquid-filled "wells." And in place of an electronic central processor, it made calculations using DNA molecules.

MAYA-II was not the first DNA computer. The idea of making such a machine was originally proposed by Leonard Adelman, a professor of computer science at the University of Southern California. In the 1994 article "Molecular Computation of Solutions to Combinatorial Problems," published in *Science*, Adelman suggested that the binding between complementary DNA strands could be used to solve computational problems. He then constructed a simple DNA computer and tested it with a "traveling salesman" problem in which a machine is challenged to find the shortest route linking seven cities. The experiment was a proof of principle; Adelman predicted that computing with molecules had enormous potential. "One can imagine the eventual emergence of a general purpose computer consisting of nothing more than a single macromolecule conjugated to a ribosomelike collection of enzymes that act on it," he wrote. By 2002 Adelman had designed a much more complex DNA computer capable of handling a problem with more than 1 million solutions. By that time many other laboratories had jumped into the field, including the team that built MAYA-II.

The tick-tack-toe game makes it easy to understand how the machine works. Samples of DNA are placed in a three-by-three grid of wells on the tray. MAYA-II gets the first move, taking the center well. When it is time for the human player to move, he or she picks a sample of DNA representing a chosen square and adds the sample to each of the wells in the grid. The sample mixes with the DNA already in the wells until it finds the "answer," a complementary sequence that it binds to. This triggers part of the strand to work as an enzyme that changes another strand of DNA, giving a fluorescent "readout" in one of the wells, which announces the computer's next move.

The enzyme reaction turns DNA strands into *logic gates,* a computer term for a mechanism that takes different types of input, performs a logical operation on them, and produces unique outputs. This is an advance from the earliest DNA computing experiments that could only yield "yes/no" answers. Solving sophisticated problems required turning DNA molecules into logic gates, which could handle more types of input and give more readouts.

Biological computers are unlikely to replace silicon-based machines anytime in the near future, but they are of great interest to researchers. They have several potential advantages over electronic machines. First, existing computers are so powerful because they are incredibly fast and can be made even faster by linking several chips or machines to carry out "parallel processing." But they still perform calculations sequentially, doing one thing and then going on to the next. In contrast, even a tiny well in a biological computer can contain billions of molecules that can carry out billions of parallel processing tasks simultaneously. A DNA computer built in 2003 by researchers at the Weizmann Institute of Science in Rehovot, Israel,

(continues)

(continued)

performed 330 trillion operations per second—more than 100,000 times faster than a personal computer.

Another interest in "wet" computing involves memory and storage. While technological advances have dramatically reduced the size of computers and increased the amount of information that can be stored in a small space, miniaturization will eventually reach a physical limit. Biological molecules also have limitations in the amount of information they can store, but this figure is likely to be thousands of times higher. The DNA in a single human cell contains nearly one gigabyte of information that can be packed into a nucleus with a diameter of about 11 micrometers (one-thousandth of a millimeter, or about 39.37×10^{-6} inches)—and DNA occupies only a part of this space. Today's standard, single-sided DVDs have a diameter about 10,000 times larger than a cell nucleus, but contain only about four times this much information.

Another motivation behind the creation of DNA computers is the possibility of linking them to cells and other living systems. In 2004 Yaakov Benenson, Ehud Shapiro, and their colleagues at the Weizmann Institute created a biological computer that could analyze levels of mRNAs produced by cells and, based on its findings, produce another molecule that could alter the activity of genes. Potentially, such a machine could be inserted into the body to look for signs of cancer and then cause the release of therapeutic molecules. In other words, a DNA computer could act as a miniaturized diagnostician and therapist.

stripped away. Any histones that remain may still bear their chemical tags. This may affect the other types of histones they bind to as the nucleus reassembles. Consequently, when daughter cells are born, they may come with "preset" patterns of active and inactive genes.

THE STRUCTURE AND FUNCTIONS OF RNAs

Originally, messenger RNA was seen mainly as an information molecule, a rather passive vehicle that carried genetic information from genes in the nucleus to the cytoplasm, where it would be made into proteins. However, scientists have since learned that RNAs play an active role in a wide range of processes in the cell. These functions arise from RNA's structure and how it interacts with other molecules.

RNA is made of nucleotides, like DNA, but there are some differences. The *R* in its name comes from the form of sugar, ribose, that it contains, as opposed to the deoxyribose sugar found in DNA. (The difference lies in that fact that ribose contains one more atom of oxygen than deoxyribose.) Like DNA, RNA is spelled in a chemical alphabet of four letters. The building blocks adenine, cytosine, and guanine are the same in both molecules, but in place of thymine, RNAs have a subunit called "uracil."

DNA forms a double helix because the cell builds long strands made of complementary bases whose structures allow them to snap together. RNA usually remains single stranded because the cell does not normally build a complementary strand. (Some important exceptions are introduced in the next section.)

Chemical interactions between the bases in a single RNA molecule cause it to fold into small secondary structures, as described for proteins in the first chapter. These shapes determine which proteins can dock on, and what happens, to the RNA.

When an mRNA is first made, it is called a pre-mRNA and contains all of the information in a gene, including protein-encoding exons and noncoding intron sequences. The molecule is immediately spotted by small machines made of other RNAs and proteins that recognize the borderlines between exons and introns and dock there. They attract another machine made of proteins and RNA—the spliceosome—which cuts out introns and glues the broken ends of exons together. This is also the stage at which alternative splicing can produce different versions of the mRNA with unique sets of exons.

The cell places a marker, an *exon junction complex* (EJC), at the sites where splicing takes place. This small group of proteins plays an important role when the mRNA leaves the nucleus. One of its jobs is to help the cell recognize RNAs made from defective genes or molecules that have not been spliced correctly.

An mRNA undergoes more modifications before leaving the nucleus. It contains information at the beginning and end that will not be translated into protein. The region at the head is called the "five-prime untranslated region," or 5′ UTR, and the tail is the 3′ UTR. ("Five" and "three" have to do with the chemistry of RNA bases and are simply used as a way to keep the directions straight.) At the head of the RNA the cell usually adds a *G cap,* a special version of the nucleotide guanine. It has two functions: It protects the RNA from being broken down by enzymes as it is being moved, and later it helps the molecule bind to the protein-making machine. The tail also undergoes changes. An RNA-cutting enzyme scans the molecule and recognizes a particular sequence of bases with the spelling *AAUAAA.* At this site the enzyme slices off the end of the RNA. Another enzyme comes and adds a huge string to the end, a long structure called the *poly A tail* because it is made up of only the base adenine, usually 100 to 300 nucleotides long.

If everything has gone well, the mRNA is complete and now attaches to a group of proteins that help escort it out of the nucleus. It passes through pores in the nuclear membrane and enters the cytoplasm. Here it is found and translated into protein by ribosomes, huge machines made of proteins and other RNAs. The components of the machine dock onto the RNA and read the three-letter codons that spell out each amino acid.

At this stage another type of molecule called *transfer RNA* (tRNA) comes into play. These molecules function as "adaptor plugs": One end is loaded with a particular amino acid, and the other side has a structure that recognizes one three-letter code of nucleotides. The cell contains many types of tRNA—at least one for each amino acid. Using a four-letter alphabet, it is possible to make 64 different three-letter codons—more than are necessary. Some of the amino acids can be spelled in several dif-

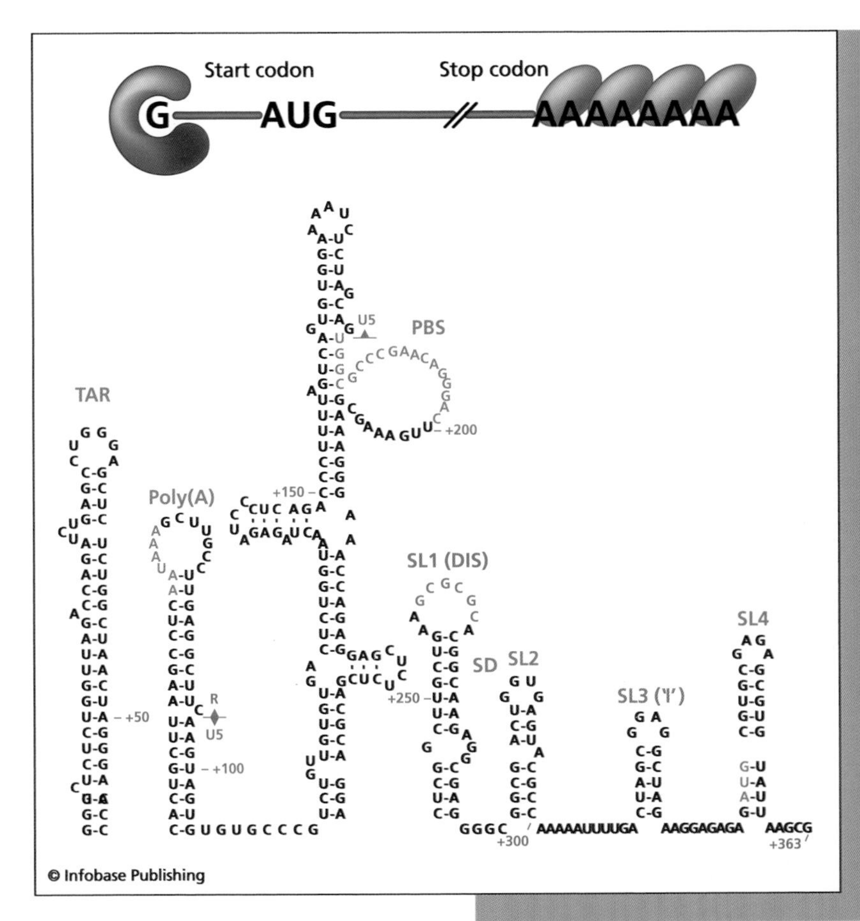

ferent ways; others are stop *codons,* whose only function is to tell the ribosome where the protein ends.

Ribosomes are huge machines made of two major subunits that come together when a protein has to be made. One of the parts, the large subunit, is made of three ribosomal RNA (rRNA) molecules and about 45 proteins. Its partner, the small subunit, has one rRNA and 33 proteins.

The top diagram of a messenger RNA molecule shows the G cap, a start signal (often the sequence *AUG*) that tells the protein-making machinery where to start working, followed by the coding sequence, the stop codon telling where it ends, and the poly A tail. If the mRNA has been spliced, groups of proteins called "exon junction complexes" are attached to the molecule. In the bottom diagram, interactions between the nucleotides that make up RNAs cause the molecules to fold in intricate structures that influence which proteins dock on and how the RNAs are used. These particular forms are found in RNAs from HIV.

They can bind to any mRNA molecule to make new proteins. The process begins when the rRNA of the small subunit docks onto a sequence that is complementary to part of the mRNA. The small subunit docks onto the mRNA first, along with the first tRNA. Then the large subunit joins, binding to a group of proteins assembled at the cap structure of the mRNA. The large subunit has docking sites where new tRNAs are plugged in, delivering the next amino acids that are needed.

RNAS AND THE CONTROL OF TRANSLATION

When mRNAs were first discovered in the 1960s, scientists believed that once created by the cell, they were almost automatically translated into proteins in the cytoplasm. This implied that the cell's population of proteins was mainly controlled by switching genes on and off. Today, mRNAs are known to play an important role in the process. Some mRNAs are destroyed before they can be used; others are stored until they are needed, or until they reach particular locations in the cell.

One way of putting an mRNA "on hold" is to attach something that hinders the activity of the ribosome. The head region of some RNAs folds into a shape that allows proteins to dock; they prevent the ribosome from attaching itself. One situation in which this happens involves the body's control of iron. The body needs *ions* (charged atoms) of this metal for several reasons; the most important is that hemoglobin molecules use them to capture oxygen and transport it through the bloodstream. But humans absorb very little iron from their diet; most is collected by recycling metal that is already in the system. The amounts must be carefully regulated because it is dangerous to have too much of the element circulating in the blood.

A protein called ferritin plays an important role in this system because it locks up iron inside cells and keeps it there until it is needed. Huge, ball-shaped complexes are assembled from 24 copies of the protein; each ball traps about 4,500 iron ions in a crystal-like structure. The cell controls how much iron is captured this way through the number of ferritin proteins it makes. But it does not do so by switching on and off the gene—that might take too long. Instead, cells make ferritin RNAs and then put them on hold until they are needed. The structure of the RNA allows this to happen. A section of the untranslated head region of the mRNA folds into a shape called an iron-responsive element (IRE). The form and chemistry of this shape allows proteins called iron-response proteins (IRPs) to bind there. They stop the translation of ferritin by blocking the ribosome's access to the mRNA strand.

Cells routinely take mRNAs apart as a way of keeping them from making too many copies of proteins, so mRNAs in storage also need to be preserved. One mechanism involves IREs in the untranslated 3′ tail regions of mRNAs. IRPs also dock there but with a different result; they prevent the molecule from being destroyed. The enzymes that take mRNAs apart usually start at the tail of the molecule and chew their way toward the head. Because mRNAs have a long poly A tail, enzymes often have to chew a long way before damaging the protein-encoding part of the RNA. (The length of the tail acts as a sort of timer.) If an IRP is bound to the tail, the enzymes have a hard time getting by, and the mRNA survives longer.

IREs were described in the 1990s by the laboratory of Matthias Hentze, a physician-turned-researcher at the European Molecular Biology Laboratory in Heidelberg. At the time this was one of the only known cases of cells blocking the production of a protein by interfering with its mRNA. Today, many similar types of control are known. Several of them involve stop codons.

At the end of the protein-encoding part of an mRNA, the ribosome reaches one of the sequences *UAA, UAG,* or *UGA* and breaks off translation. Mutations or copying errors in DNA often lead to stop codons in the wrong places. Because it takes three letters of the DNA and RNA alphabet to encode one amino acid, adding or losing three letters (or a multiple of three) may not cause too many problems. Proteins often still function if they lose or gain a few amino acids. But if the mistake affects a number of letters not divisible by three, the result is a *frameshift:* Suddenly all of a molecule's codons are misspelled. This is a serious problem because the ribosome reads an RNA's sequence in groups of three to pick the right tRNA and the right amino acid. After the error it will build a completely different molecule until it reaches a stop codon.

The situation is like reading a sentence of three-letter words: "The mad old dog ate the hat." Adding or removing three letters may still produce a readable sentence: "The old dog ate the hat." But, if a single letter is added or removed, all the boundaries between words shift: "Thh ema dol ddo gat eth eha t." This looks like complete nonsense unless one knows what has happened, and the same is true in the cell. A frameshift causes the construction of a completely different protein. Since there are only 64 possible combinations of letters, and three of them are stop codons, mRNAs with frameshifts usually contain a stop codon much too early in the molecule.

Many of these mRNAs are caught by the cell and destroyed in a process called *nonsense-mediated mRNA decay,* or NMD. This mechanism notices defective RNAs and adds a protein tag that causes them to be attacked by RNA-digesting enzymes. One way that NMD recognizes improper stop codons has to do with splicing.

When it comes time for translation, an mRNA still contains markers that it has been spliced—the EJCs, described in the previous section. They do not pose an obstacle to translation because ribosomes can simply shove them away. There is one important exception. RNAs are rarely spliced near the end, which means that in a healthy molecule, an EJC should not be too close to a stop codon. If this is the case, however, the molecule is recognized by NMD. The process of translation is interrupted, and other molecules come to carry the RNA away and break it down.

Like most processes in the cell, NMD is extremely important when it functions correctly but can cause new problems if it recognizes the wrong molecules. Usually it leads to the total destruction of defective RNAs—even when it would be better to have a flawed version of a protein than none at all. In 1989 the laboratory of Lynne Maquat, a biochemist at the Roswell

Park Memorial Institute in New York, showed that NMD contributes to beta-thalassemia, the most common disease in the Western world that is caused by a single gene. Beta-thalassemia reduces the body's production of hemoglobin, which is needed to carry oxygen through the blood. The disease arises in people who inherit a mutant form of a gene called beta-globin. The error creates beta-globin with a stop codon too close to an EJC. When NMD breaks it down, the body loses an important molecule. In this case, an intended safety mechanism is actually attacking the body.

Until recently, NMD was thought to spring into action only to control defects, but now it is recognized as a mechanism that cells use widely to control the quantity and quality of many molecules. Alternative splicing sometimes produces RNAs with nonsense codons; for some reason errors in the cut-and-paste operation produce a bit of nonsense code in the middle of an mRNA. In 2004 R. Tyler Hillman, Richard Green, and Steven Brenner of the University of California, Berkeley, carried out a computer analysis that showed that about one-third of the time, alternative splicing places a stop codon too close to a splice site. This activates NMD, which eliminates most of the RNA before it can be transformed into proteins.

The same year the laboratory of Harry Dietz, a physician and researcher at Johns Hopkins University School of Medicine in Maryland, studied this effect in the cells of mammals. They shut down the NMD machinery by removing one of its most important components, a protein called Upf1. This changed the behavior of a huge number of genes: About 10 percent of the genes they studied became more productive, probably because spliced forms that normally would have been caught by NMD and destroyed were slipping through.

This high number of splicing errors is now thought to have an important function. In some cases the cell might not be able to block the activation of a gene, and it may produce too many mRNAs or make them at the wrong times. By splicing these molecules incorrectly, the cell creates mRNAs that will be destroyed through NMD. This gives it another chance to block the production of a protein at a later stage.

NONCODING RNAs

In 2000, as the human genome was nearing completion, it appeared that less than 2 percent of the sequence encoded proteins. A more detailed version of the genome, published in the October 21, 2004, edition of *Nature,* confirmed the findings and reported that human cells likely encoded only between 20,000 and 25,000 genes. For most scientists the small number of genes came as a great surprise. Ewan Birney, who analyzes genomes at the European Bioinformatics Institute in Hinxton, Great Britain, had been collecting scientists' guesses in a contest called "Genesweep." A few scientists came close with estimates in the 25,000 range, but most guessed much higher; the average prediction was 60,000. Did the other 98 percent of human DNA have a function, or was it just junk, a sort of "excess baggage" of evolution? A partial answer is now emerging with the discovery that cells build a huge number of *noncoding RNA* molecules.

Many of these RNAs had been found in experiments but were often interpreted as artifacts, or fragments of mRNAs on their way to be translated into proteins. Then in 2004 Eric Schadt, Stephen Edwards, and their colleagues at the Rosetta Inpharmatics company in New York, working with scientists at several other companies and the Scripps Research Institute in Florida, decided to carry out a new kind of experiment. They used a technology called the DNA chip (also known as the DNA microarray) to take a much closer look at the RNAs produced by cells.

DNA chips were developed in 1994 by molecular biologist Pat Brown of Stanford University and the California-based company Affymetrix. The technology acts as a surveillance system that can detect whether cells have produced RNAs from genes or other DNA sequences. It is based on the principle that complementary molecules bind to each other. Scientists extract all the RNAs from a cell and pass them over a series of probes representing molecules that are complementary to genes. If a cell has made RNA from one of the genes, it binds to the probe and gives off a signal. If no RNA has been made, there is no docking and no signal. The most common use of this technology has been to compare the gene activity of different types of cells—for example, a tumor cell and a similar but healthy one. Scientists hope that the method will reveal genes that behave strangely during cancer—perhaps because they have caused the disease. At the very least, identifying such molecules could help diagnose cancers at an early stage. Detecting certain molecules in the body could provide an early-warning system that signals the presence of cancer even before a tumor forms.

Making a DNA chip requires assembling a set of probes. Previously, the difficulty and expense of doing this—and scientists' assumptions about what they were likely to find—meant that most laboratories did this by picking known genes or sequences that had been identified by computer as probably belonging to genes. Schadt and his colleagues decided to try something different. They used probes for known human genes but added samples of all the DNA sequences from two entire chromosomes. This new type of chip, called a "tiling array," made no assumptions about what part of the genome was transcribed into RNA. It simply asked two entire chromosomes which parts were being used. Making the tiling array was a huge task, involving the creation of more than 3.7 million different molecular probes. They could scan for even very tiny RNAs that might have been missed using long probes for single genes.

The results were stunning. First, Schadt showed that the probes were accurate; they detected more than 95 percent of the exons already known to exist on the two chromosomes. But they also revealed about 3,000 RNAs unconnected to any known gene. Many were transcribed from the "second strand," the DNA sequence opposite a known gene. Schadt and his colleagues concluded that about one-fourth of these might represent genes that had gone undetected; they did not know the functions of the rest.

As other groups began to use tiling arrays, they started to get an idea of the number of such "mystery RNAs" produced by cells. In 2005 Jill Cheng, Philipp Kapranov, and their colleagues at Affymetrix put together a tiling array for 10 human chromosomes, containing more than 74 million probes, covering about 30 percent of the human genome. (The 23 human chromosome pairs do not all contain the same amount of DNA or the same

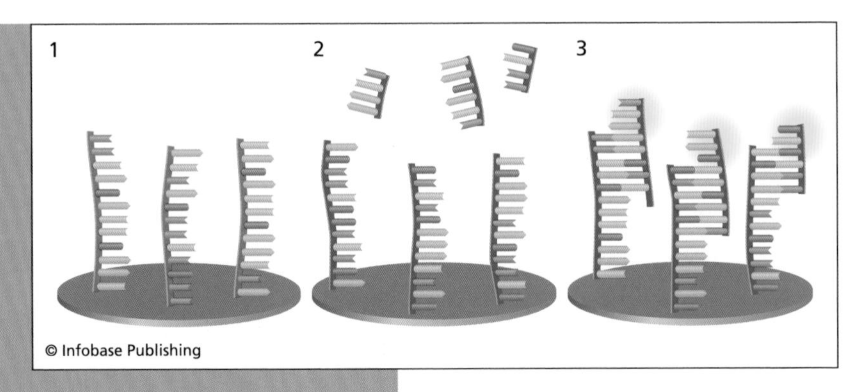

DNA chips, or DNA microarrays, consist of probes made of DNA. The method is based on the fact that complementary sequences of nucleotides bind to each other. First, Probes are assembled on a chip (A). Next, RNA molecules are extracted from cells, prepared, and washed over the chip (B). If the cell has made an RNA from the gene on the probe, it binds and gives off a fluorescent signal (C). By attaching different fluorescent markers to samples from two cells, scientists can compare the gene activity of, for example, healthy and cancerous cells.

number of genes.) They discovered that on the average, about 10 percent of each chromosome is transcribed into RNA. This is between five and 10 times the amount of RNA found in exons.

Other species use an even higher percentage of their genomes to produce RNA. In another project from 2005, two Japanese groups—the RIKEN Genome Exploration Research Group and the FANTOM Consortium—examined the complete mouse genome and discovered that at least 62 percent of the sequence was transcribed. This was more than double the amount known to make up genes. Scientists now believe that most of these molecules are RNAs that do not encode proteins. They have other functions that experiments are beginning to reveal.

One discovery has been that cells make a huge number of small RNAs, *microRNAs,* whose job is to control the use of mRNAs. MicroRNAs do this because their sequences are complementary to existing mRNAs. This allows the two types of molecules to bind to each other, creating a double-stranded RNA molecule.

Messenger RNAs made by the cell are not normally double stranded, as pointed out in the previous sections, but viruses often carry double-stranded RNA. The cell recognizes these molecules as foreign and destroys them. The same thing happens to mRNAs that have been made double stranded through the attachment of a microRNA, even when it has only docked onto a small region of the molecule. MicroRNAs can also act in a second way: Like the IRPs introduced in the previous section, they can sit on an mRNA and interfere with its translation.

In 2008 Nikolaus Rajewsky, a researcher at the Max Delbrück Center for Molecular Medicine in Berlin, found a clever way to distinguish between these two types of activity. "Experiments using DNA chips have shown us how the population of many mRNAs drop when the cell overproduces a certain microRNA," Rajewsky said in an interview with the author. "But these experiments did not reveal how proteins behaved, so we really have not had a clear idea of how important this type of control is to the cell." Matthias Selbach and other members of the lab found a way to grow cells in a medium containing "heavy" versions of the amino acids needed to build proteins. (Some of the atoms in the amino acids were isotopes—they had extra neutrons.) The researchers prompted cells to make higher-than-normal amounts of microRNAs and then compared the proteins they made to those of normal cells using mass spectrometry. The "heavy" amino acids allowed the scientists to tell the difference between molecules made by the control cells and those with high levels of microRNAs. "We discovered that a single microRNA can lower the production of hundreds of different proteins," Rajewsky explained. "But it acts more like a volume control than an on/off switch. Rather than completely eliminating the protein, a microRNA usually tunes down the amount that is made, to the point that the cell may only make about a fourth of the normal amount."

In 1998 biologists Andrew Fire of the Carnegie Institution for Science in Baltimore, Maryland, and Craig Mello of the University of Massachusetts Medical School turned the "interference" between RNA molecules into a new method of controlling genes in laboratory animals. The strategy was to create artificial RNAs, called *small interfering RNAs* (siRNAs), with sequences that are complementary to an mRNA. Put into cells, these molecules dock onto other RNAs and block the production of their proteins. Multiple genes can be shut down as well because several siRNAs can be introduced into cells at the same time. Almost overnight, Fire and Mello had produced a powerful new method to control and study genes. Their reward was the 2006 Nobel Prize in physiology or medicine. Nick Hastie, professor of human genetics at the Medical Research Council in Great Britain, summed up the importance of the work in this way: "It is very unusual for a piece of work to completely revolutionise the whole way we think about biological processes and regulation, but this has opened up a whole new field in biology."

Another advantage of the method is that it can be used in adult animals. It is not necessary to engineer the organism's genome ahead of time, which means the strategy might also be transformed into a treatment for diseases. Theoretically it is possible to create an siRNA to match any cellular molecule, including one that has malfunctioned and caused disease. An siRNA might shut it down, if the molecule can be delivered to the right types of cells. This approach has already been used, with promising results, in human clinical trials as a treatment for an eye disease called macular degeneration.

By controlling the output of genes, microRNAs are another "loophole" in the central dogma of molecular biology, "DNA makes RNA makes proteins." The central dogma is still true as an overall outline, but at every step along the way—from the activation of a gene to the construction of a spliced mRNA to its translation into protein—scientists have discovered new mechanisms by which the cell can block the process. This likely has to do with the amount of energy needed to transcribe genes and the overall difficulty of doing so. If the cell makes lots of mRNAs it has them on hand when they are needed quickly. On the other hand, if it does not need them, it has several methods to prevent their translation into proteins. The alternative would be to make what is needed only on demand, but activating a gene is a complex process that often involves making several other molecules, changing the positions of nucleosomes on the DNA strand, and other operations.

The cell contains several other types of noncoding RNAs. The following table provides an overview.

A SHORT CATALOG OF RNAs	
Name	Function
MicroRNA	Docks onto mRNAs to cause their breakdown before they are translated
Ribosomal RNA (rRNA)	Forms the core of the machine that translates mRNA into proteins
Signal recognition particle RNA	Helps move proteins to membranes as they are translated
Small interfering RNA	Behaves like microRNA but is artificially made
Small nuclear RNA	Assists in splicing and other processes involving RNA and DNA in the nucleus
Small nuclear RNA	Makes chemical changes to rRNA and tRNA
Transfer RNA (tRNA)	Functions as "adaptor plug" to match amino acids with a corresponding codon
Y RNA	Plays a role in the handling of misfolded RNAs and the copying of DNA
These are the main types of RNAs and their functions. The cell produces several other types of RNA molecules that are not used to produce proteins.	

TRANSPOSONS

Despite the fact that much of the noncoding material in the genome appears to have functions—it is likely used to produce microRNAs and other types of control elements—there is still a lot of "junk" in human DNA. Another great surprise from the analysis of the genome was that a huge amount of it seemed to be made of small bits of information—usually about 300 bases long—repeated over and over again. These elements are called "Alu sequences," or "Alu repeats" (the name comes from *Arthrobacter luteus,* a bacterium in which the sequence was first identified). Scientists estimate that there may be 1 million copies in the genome, which would account for more than 10 percent of the entire human sequence. They are scattered everywhere, almost as if someone copied a sequence hundreds of thousands of times and pasted it in again at random places. In fact, this scenario is probably not far from what actually happened.

The study of retroviruses such as HIV, which causes AIDS (acquired immunodeficiency syndrome), has given researchers some insights into how Alu sequences might have entered the genome and become so widespread. Retroviruses also introduce new genes into the genomes of cells. HIV brings along an entire toolbox of RNAs and proteins that allow it to reproduce. Like most viruses, it carries only what it needs, taking advantage of molecules in the cells it infects to survive and reproduce. For instance, instead of carrying along molecules to transcribe its own genetic information (which is made of RNA), HIV borrows them from the cell.

Upon entry into the cell the virus is taken apart, and its RNA and some of its proteins enter the nucleus. One of these molecules is called a "reverse transcriptase," a tool that carries out transcription in reverse. Instead of reading DNA and making a single-stranded RNA, it reads RNA and assembles a single strand of DNA. Another protein tool stitches this strand into the cell's genome. Once there, it is treated like any other human gene; the cell transcribes it into RNA that is then transported to the cytoplasm. Some of the RNA is translated into proteins that the virus needs. These are then collected with RNAs and packed into membranes to make new copies of the virus.

Reverse transcriptases were independently discovered by two researchers in 1970: Howard Temin (1934–94) of the University of Wisconsin–Madison and David Baltimore (1938–) at the Massachussetts Institute of Technology. The discovery was dramatic because it obviously contradicted the central dogma's position that information could only flow from DNA to RNA and not vice-versa. It also helped explain the behavior of viruses that had been linked to the development of cancer. Both Temin and Baltimore had previously worked on viruses and their reproduction in the laboratory of Renato Dulbecco (1914–) at the California Institute of Technology (Dulbecco later moved to the Salk Institute), and for their discoveries the three men were awarded the 1975 Nobel Prize in physiology or medicine.

Some species contain genes that behave in a similar way. They begin as genes that are transcribed into RNA; the RNA is then reverse transcribed into DNA and reinserted somewhere else. Such molecules are called *transposons,* or "jumping genes," for their ability to jump from place to place in chromosomes. This is in contrast to normal genes, which usually occupy fixed locations.

Barbara McClintock's pioneering studies of heredity in maize revealed aspects of gene behavior that would not be understood or accepted for nearly 50 years. Here, she is shown at Cold Spring Harbor Laboratory in the 1930s. *(Barbara McClintock Papers, American Philosophical Society)*

Geneticist Barbara McClintock (1902–92) discovered transposons in the 1930s and 1940s while working on maize at the University of Missouri and at Cold Spring Harbor Laboratory, in New York. She was curious about how genes directed the color of kernels on corncobs. Patterns of dark kernels were encoded in genes, but the inheritance of the patterns could not be accurately predicted using Mendel's laws of heredity, as they did not follow normal patterns. McClintock wondered why this was happening, so she began studying the process in cells. As she examined maize chromosomes under the microscope, she observed regions that seemed to have jumped from one position to another as DNA was reproduced. She called them "jumping genes," or transposons. Although her observations were accurate and her hypothesis explained the facts, the idea that genes could behave this way was too radical at the time. Few scientists understood or believed her findings. The discovery of reverse transcription, however, opened the door to the idea that cells might contain their own mechanisms for converting RNA into DNA, and McClintock's work was finally appreciated—half a century after she had first proposed the idea. She was awarded the Nobel Prize in physiology or medicine in 1983.

Transposons offer an explanation for Alu sequences and a great deal of the other "junk" in genomes. Researchers believe that these sequences originated as a virus that infected an ancient primate. The virus wrote its RNA into the DNA of the host's cells. At some point the infection reached sperm or egg cells, and the new gene entered the species's hereditary material and became part of the primate genome. It spread as RNA was transcribed from the gene in each new animal. The RNA contained the tools to make another DNA strand and insert the new copy into another place in the genome. Whenever that happened in a reproductive cell, it produced more Alu sequences in the genome.

Two types of transposons are known. *Retrotransposons* function like retroviruses. *DNA transposons* work a bit differently. Once in an organism's genome, the gene is also used to make new RNAs and proteins, but instead of reverse transcribing the RNA, the molecules encoded in the transposon physically cut their own gene out of the genome and move it to another place in the sequence.

The Alu sequence and other jumping genes have been a major factor in evolution. First, they have dramatically increased the size of plant and animal genomes, giving evolution fresh new material to work with. Researchers have estimated that up to 80 percent of the genome of maize and nearly 90 percent of the wheat sequence originally came from transposons. Their artifacts make up almost half of the genome of humans and other animals. Over long periods of time mutations can change the new sequences and make them useful to an organism. It is estimated that about 100 human genes have been "domesticated" in this way. The gene Harbi1, which probably plays a role in repairing DNA, was acquired from a transposon called Harbinger.

Such domestications are nevertheless relatively rare. Transposons are usually dangerous because eventually, inserting DNA sequences into random places in the genome causes problems. Sometimes they land in the middle of another gene, which nearly always destroys it.

As far as scientists know, all of the transposons in humans, mammals, and other vertebrates have become inactive through evolution except for the domesticated genes, which have taken on other jobs and no longer jump. Deactivation happens because transposons suffer random mutations, just like the rest of the genome. This often destroys the tools' ability to cut, copy, or paste themselves. That usually happens fast; while natural selection protects most genes from severe changes, disarming a transposon usually helps a species. A gene that can no longer jump is no longer a threat to other genes. Eventually every transposon is likely to become extinct for this reason, unless it finds an "escape route." For example, it might move to another species by catching a ride on bacteria or viruses. A study of genomes reveals that this has happened many times; closely related transposons are found in a wide range of species.

Active transposons are like extremely simple genetic engineering toolkits that add sequences to an organism's DNA. This has led scientists to consider using them as vehicles to deliver healthy genes to people who have genetic diseases. About 5 percent of the human population has inherited a disease linked to a defect in a single gene. And many other illnesses, including some forms of cancer and most "old age" diseases such as Alzheimer's, have clear genetic components. The symptoms of some of these diseases can be treated, but drugs do not address the underlying cause—a flaw in the genetic code. A real cure would involve giving a person a healthy version of one or more genes. That requires a delivery vehicle, able to slip past defenses that normally protect cells from foreign DNA or other molecules. A common approach to "gene therapy" is to build healthy genes into viruses that can evade the defenses. This strategy has been tested in a number of clinical trials, but there have been side effects, sometimes due to the very complex and unpredictable nature of viruses. It will likely take many more years for the problems of viral therapies to be worked out.

In the meantime, Zoltán Ivics, a world-renowned transposon researcher at the Max Delbrück Center for Molecular Medicine in Berlin, thinks that these jumping bits of DNA could provide a shortcut. Transposons are small, which should make it relatively easy to insert them into a cell. The transposon would be rebuilt so that it would not be dangerous. It would not contain the tools needed to jump to other places or to infect germ cells.

The first step in making such a molecule is to start with a version that works in animals' cells, where it can be tested, but all of the existing transposons in vertebrates have been inactivated through mutations. In the early 1990s Ivics, his wife, Zsuzsana Izsvacs, and other members of Perry Hackett's laboratory at the University of Minnesota decided to try to "revive" a transposon by repairing the mutations in an ancient gene called Minos. They began by comparing damaged copies of the gene in species where it had been most recently active: salmon and other fish. The researchers picked 12 copies that seemed to be the most intact. The comparison revealed the nucleotide "letter" that occurred most often in each position of the gene, which was most likely to be the original. When the scientists made a new molecule with the corrected spellings, it gave them a new, artificial transposon that functioned again. They dubbed it "sleeping beauty"—as it had been "sleeping" in the fish genome a very long time—and began testing whether it could be used to transfer genes into laboratory cultures containing human cells. When the cells grew, Ivics and his colleagues found that sleeping beauty had been inserted into their genomes. The gene had been delivered, and it now seemed to be able to copy itself and jump to new positions in the chromosomes.

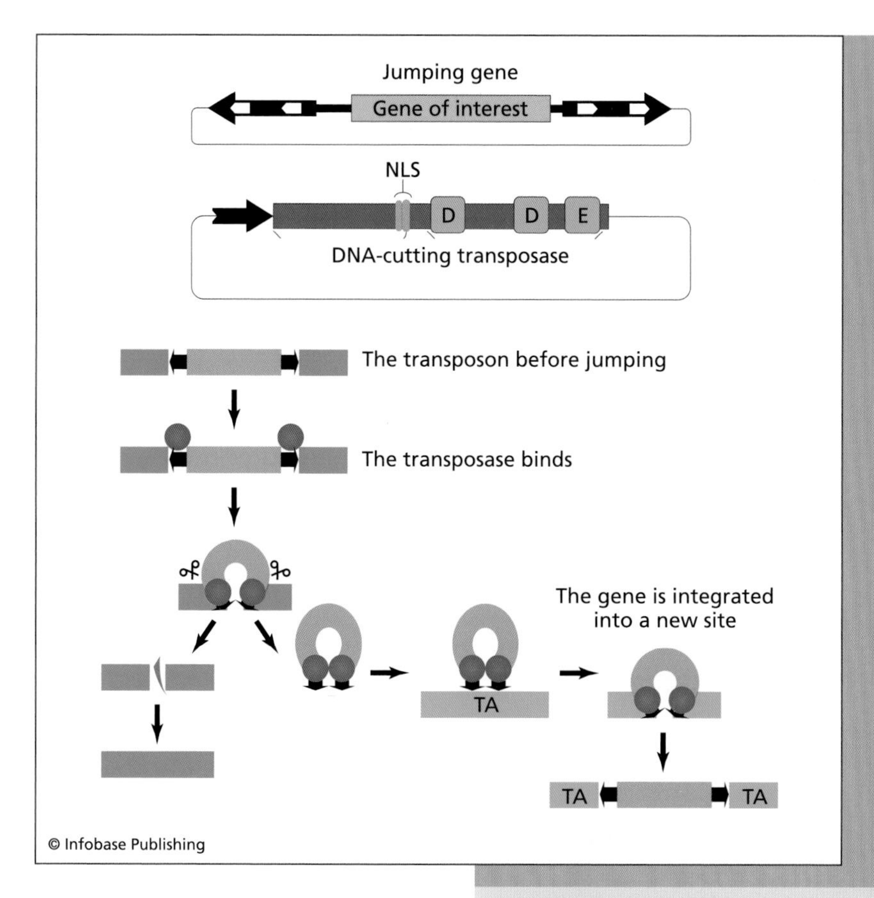

How DNA transposons jump: The gene (orange) is located between two sequences recognized by a transposase (purple). The transposase binds and cuts out the gene. This leaves broken ends of DNA that are mended by other molecules. Now the transposase recognizes and docks onto target sequences (green) and inserts the gene in a new location.

Using transposons for gene therapy will require not only getting new genes into cells but also ensuring that they behave properly once inside. A clinical trial conducted in hospitals around the world in 2002 and 2003 showed how crucial this is. The study involved a virus rather than a transposon. Researchers had used a retrovirus to deliver a healthy version of the interleukin 2 receptor protein to patients suffering from a severe immune system disease called X-linked severe combined immunodeficiency, or X-SCID. People with defective versions of this molecule are unable to produce B and T cells, white blood cells that play a key role in the immune system. The lack of these cells allows even trivial infections to run rampant in a person's body. People suffering from X-SCID have to spend their life in a completely sterile "bubble" unless a bone marrow transplant from a close relative takes hold. Even then, the therapy usually has only temporary effects, so it has to be repeated many times. Gene therapy might be the way to give patients a relatively normal life.

But in several patients who participated in the trial, the virus inserted the gene's DNA into the worst possible position. It landed near the control region of a gene called LMO2. This molecule is a proto-*oncogene,* which means that it is known to cause cancer if it undergoes changes. The presence of the new gene had a similar effect. Somehow it triggered an uncontrollable wave of replication among T cells, leading to a leukemia-like form of cancer.

The only way to prevent such effects, even with transposons, is to control where the gene lands. In the best case a scientist could program a transposon to insert itself into precise places in the genome by rewriting the tools that tell it where to go. In June 2007 Ivics and Izsvacs used sleeping beauty to demonstrate that this could be done. They now hope to rebuild the molecule to carry extra therapeutic genes and to reprogram it to insert itself in specific places.

MAKING AND REFINING PROTEINS

The first chapter introduced the main features of protein structures and their investigation using X-rays and other methods. Proteins are folded into small secondary structures and larger domains—the shapes that determine what other molecules they bind to and how they behave—as they are translated from RNA. Then they undergo more changes that affect their functions.

There are two main types of *translation.* One method is used for proteins that will be secreted into cells, or inserted into membranes (such as receptors). Another is used for proteins that work inside the cell but are not embedded in membranes.

Both types of translation begin when ribosomes attach themselves to an RNA, read its three-letter nucleotide codons, and link up to a RNA that bears the right amino acid. What happens next is crucial in delivering molecules to their duty stations. The chemical environments of the membrane and the cell exterior are quite different from that inside, so proteins that work in these places have special chemical properties. Such molecules are not suited to work inside the cell, so as they travel outward, they are packed into compartments made of membranes, called *vesicles*.

Packaging begins during the process of translation. Early in the process of reading an mRNA, the ribosome discovers a signal—a particular sequence of amino acids—that tells it to stop and deliver the molecule to a set of saclike membranes called the "endoplasmic reticulum" (ER). There, the mRNA is brought into contact with a protein called the "signal recognition particle" and a receptor on the ER membrane. The receptor has a channel to the inside of the ER, and the tip of the protein is pushed through, like a needle pulling a thread through cloth. This tells the ribosome to start translating again, and as it does so, it pushes the rest of the protein strand into the ER. The protein is passed through the ER, and then other signals tell the cell whether to secrete it or move it to a particular set of membranes.

Other RNAs are translated into proteins in the cytoplasm, and many contain signals that tell the cell how to handle them. The RNAs or proteins carry instructions—like postal codes—revealing a protein's destination. Histone proteins, for example, carry a pattern called a "nuclear localization signal" that directs them to the cell nucleus. Other signals guide proteins to various organelles or specific regions of the cell.

At the end of translation, a protein falls away from the ribosome. It has been folding during translation and now takes on its finished shape, often with the help of partner molecules. As described in the first chapter, the amino acids are strung into a chain (primary structure) that folds into simple secondary structures (alpha helices and beta strands) and more complex tertiary structures (domains). Folding has to be done right: In the inside of the molecule, it brings together amino acids that need to interact; it also creates outer surfaces that allow the protein to bind to other molecules.

The protein may need to undergo additional *post-translational modifications.* The most important are the following:

- Glycosylation, the addition of sugars to some of the amino acids in a protein, happens in the ER or another set of membrane sacs called the Golgi complex. Some types of modifications act as additional postal codes, giving the cell more specific information about where to put the protein. Others modify the shape and chemistry of the protein to enable it to dock onto different partner molecules. Many proteins only function properly if specific sugars are added to precise locations in the molecule.
- Proteolysis, or cutting the protein strand, occurs in cases when the protein needs to be trimmed so that it will fold in a particular way. Sometimes the cell only needs part of a molecule.
- *Phosphorylation,* or adding clusters of atoms called phosphate groups to specific amino acids, changes the shape of a protein and often influences the sites where it can bind to other molecules. Adding phosphate groups to proteins and stripping them off again is one of the main means by which information is transmitted through the cell.

SUMMARY

On the *Nova Online* Web site (http://www.pbs.org/wgbh/nova/genome/debate.html) in April 2001, reporter Kevin Davies wrote: "The most shocking surprise that emerged from the full sequence of the human genome earlier this year is that we are the proud owners of a paltry 30,000 genes—barely twice the number of a fruit fly. . . . This unexpected result led some journalists to a stunning conclusion. The seesaw struggle between our genes—nature—and the environment—nurture—had swung sharply in favor of nurture."

The number appears to be even lower than Davies suspected. With such a small number of genes, how can humans still be so complex? One lesson that has been learned from the genome and the newly discovered mechanisms of control is that even a "small" set of genes can create incredibly complex biological structures, partly because of the nuances by which the effects of those genes are regulated. MicroRNAs carry out one type of control. Others arise from the mechanism by which the information in genes is translated into proteins. Each step gives organisms a way to regulate a gene's use or to create different versions of the molecule it encodes. This significantly expands the amount of information in the genome.

3

Communication Between and Inside Cells

The previous chapter focused on DNA and RNA, leading to a main theme of the rest of the book: how proteins manage the business of the cell. Proteins have often been called the cell's "worker" molecules because they play a central role in most biological processes. This image is too simple, as the previous chapters have shown, because proteins usually work in machines containing many molecules, often including RNAs, sugars, and fat molecules. Even so, explanations for most biological processes usually begin with the identification of the proteins that are involved, studying how their architecture governs their behavior, and unraveling their interactions with other molecules.

The next chapters examine the main types of jobs carried out by proteins. A crucial part of every cell's life is sensing its environment and adjusting to external conditions. For a single-celled organism this may involve swimming or crawling to find a new source of food. In the human body the environment consists of neighbors that tell a cell how to behave and what it should become. This chapter introduces the main types of molecules involved in receiving environmental signals and passing them on to genes.

KINASE RECEPTORS AND LIGANDS

The surface of the cell is a rugged, swampy landscape interrupted by thickets of brushy receptor proteins and pitted channels formed by membrane proteins. These two structures manage a great deal of the flow of information between a cell and its environment.

While the focus here will be proteins on the cell surface, some molecules inside the cell are also considered receptors. For example the estrogen receptor alpha protein is bound to DNA in several places. When triggered by the hormone estrogen, it activates a number of genes. Estrogen and other molecules that bind to receptors are ligands. Some hormones are so small that they simply slip into the cell through the membrane, head directly for the nucleus, and pass through its membrane. But larger ligands, or molecules that carry a charge, cannot enter directly; they usually work by triggering a receptor floating in the cell membrane.

Membrane receptors have at least one domain outside the cell that comes in contact with ligands, which may be free-floating molecules or proteins on the surfaces of viruses or other cells. A second domain of the receptor passes through the two layers of the membrane. The first chapter used the image of one soap bubble inside another to introduce the double-layered cell membrane, which is formed from fat molecules called lipids. A protein's transmembrane domain is often crucial to the molecule's function. In some cases the module is as simple as a single alpha helix, but many proteins have several helices that weave in and out, like a thread that has been stitched through a cloth several times. For example, G-protein coupled receptors (GPCRs) pass through the membrane seven times, in seven alpha helices. GPCRs have many roles—including important functions in vision and the other senses—and are discussed later in the chapter in the section on "Photoreceptors and Molecular Amplifiers." Another role of transmembrane structures—often alpha helices but also beta sheets—is to create channels that permit the passage of molecules or ions through the membrane.

Many transmembrane proteins are *kinase receptors,* whose name comes from a chemical transformation they carry out. Kinases are proteins that work by transferring collections of atoms called "phosphate groups" from one protein to another. This process of transfer is phosphorylation. Phosphates act as a sort of "energy currency" in the cell. They come in various forms. *Adenosine triphosphate* (ATP) is a high-energy compound with three phosphate units. It can be broken down into ADP (adenosine diphosphate, with two units) and AMP (adenosine monophosphate, with one); when this happens, energy is released. The donor molecule usually becomes inactive. The recipient, on the other hand, is usually activated. Its shape is changed so that it can dock onto other molecules and switch them on by transferring phosphates to them.

A receptor becomes activated when a ligand binds to it. Hermann Emil Fischer (1852–1919), German chemist and winner of the 1902 Nobel Prize in chemistry, compared the interaction to the way a key fits a lock. Binding causes a change in the receptor's structure, usually the part of the molecule inside the cell. Ligands often bind to two "heads" of the receptor outside the cell simultaneously. This draws them together—like pushing together hangers in a closet. Just as the latter brings together the clothes that are hanging from them, the transmembrane domains and tails of receptors are brought together in the cell. If the tails come close enough they bind, changing shape and becoming active. They are now able to use energy from ATP to obtain phosphates and activate other molecules in the cell. Many receptors have "keyholes" to fit several ligands. Each combination has a unique effect on the structure of the receptor and changes the set of other molecules that it can bind to.

Mutations that affect receptors often have dangerous consequences. Some changes make receptors always active, even when they are not bound to ligands. Several diseases have been linked to such changes. A 2007 study by Eric Clauser's laboratory at the Cochin Institute of Descartes University in Paris, France, connected a receptor to hypertension and the hardening of arteries in the hearts of mice. The work, published in the *Journal of Clinical Investigation,* showed that these conditions

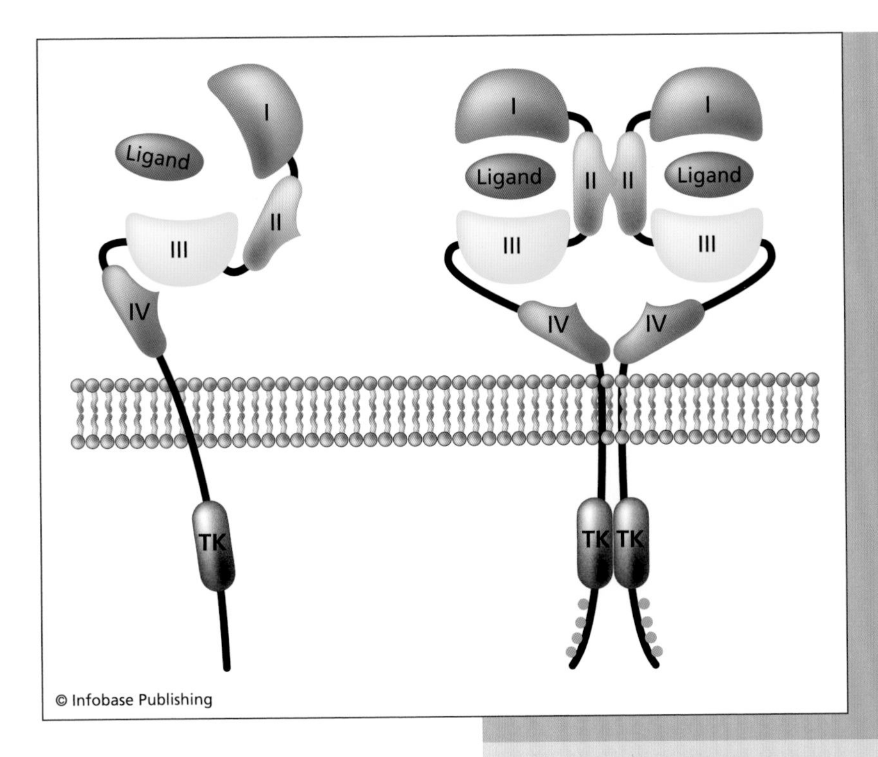

The epidermal growth factor receptor (left) is a good example of a kinase receptor. It has four extracellular domains (I-IV). Docking onto a ligand (purple) changes the structure, causing two copies of the receptor to bind to each other (right). This activates the receptor's intracellular tail, which triggers a signaling pathway that activates genes. Many receptors can bind to several ligands, changing their structures in different ways, thus activating different signaling pathways.

arise in mice that have a permanently active receptor for a molecule called angiotensin. A study the same year from the laboratory of Zhixiong Li, a hematologist at the Hannover Medical School in Germany, linked an overactive receptor for the neurotrophin protein to leukemia. Similar problems have been found in other cancers and conditions such as drug addiction.

Mutations are not the only problems that confuse receptors and their ability to send signals. If a cell produces too many copies of a receptor, it can become oversensitive to signals, which may also lead to disease. Overproduction of a receptor for a molecule called the epidermal growth factor (EGF) has been connected to cancer of the ovaries. Likewise, the production of faulty ligands—or the wrong amounts of these molecules—is a frequent cause of disease.

One of the many functions of receptors and ligands is to help cells learn how to develop. They receive cues from their neighbors that tell them where they are located in the body and what types of tissues they should become. This topic is covered in more detail in the section called "Developmental Signals and the Origins of Animals," at the end of this chapter.

Another important function of signals is to guide cell migrations. Receptors and ligands lead white blood cells through the bloodstream and lymph systems, guiding them as they leave vessels to enter particular tissues and organs. A spectacular example of pathfinding is the way the nervous system establishes connections to "wire" the brain and body. For example, nerve cells near the spine of a giraffe extend long branches called *axons* into the neck; the axon of a single cell may be more than 13 feet (4 m) long. As it grows, it searches for particular cells that it needs to bind to. If it fails, the nervous system will not be wired properly. The same thing—on a slightly smaller scale—happens during the formation of a human embryo.

Some of the most important guidance cues for axons depend on ligands called ephrins, which dock onto receptors called ephs. This encounter usually repulses the axon, telling it to grow in a different direction. At the target, different ligands and receptors tell it to stop and bind.

MEMBRANE CHANNELS AND PUMPS

Ligands and receptor proteins are one of the main ways that cells sense each other and the environment. A subset of proteins in the membrane called *membrane channels* permits cells to communicate in another way, by changing their electrical charge. These molecules also help maintain a balance between the environment of the cell and the outer world, regulating how cells take up water, energy, and nutrients.

Membranes undoubtedly played a central role in the evolution of life. Without them proteins that work together would float away from each other, and delicate interactions would be disrupted by random molecules drifting by. But cells need nutrients and energy from the environment and have to expel their wastes, so membranes cannot be impenetrable barriers. Instead, they act as border guards and customs officials, carefully screening what moves in and out. Some atoms, such as water, are small enough simply to pass through the membrane. Larger molecules and charged particles usually need help to make the passage. It is important for the cell to regulate its electrical charge, and it does so by controlling the inward and outward movement of ions, atoms, or groups of atoms bearing a charge.

This work is carried out by protein channels and *ion pumps* that can be opened or closed. If a cell's membrane channels were open all the time, negatively charged atoms (or groups of atoms) would flow into positively charged areas until the interior charge matched that of the environment. Sometimes this is the case, but often an imbalance is important, and it is controlled by channels—pores that can be opened or closed—or pumps, which do the electrical equivalent of pushing water uphill. Some pumps are also exchangers; they move different particles in opposite directions. All the cells of the body have a type of pump that pushes out positively charged sodium and draws in positively charged potassium.

The main charged particles that are moved across cell membranes are sodium, potassium, chloride, and calcium. Each relies on its own specialized set of channels and pumps, which are usually triggered to open or close by a ligand or another membrane protein. They may also directly sense changes in the balance of charges between the cell and the environment and respond by themselves.

Water normally passes freely through the membrane, but certain types of cells (for example, kidney or red blood cells) need to accumulate extra water. Peter Agre (1949–), a researcher at Johns Hopkins University, discovered the first known water channel (see sidebar). The finding was considered so important that Agre was awarded the 2003 Nobel Prize in chemistry. In the meantime, researchers have discovered at least 11 closely related types of aquaporins that help recover water from body fluids. This is particularly important in the kidney, whose cells have to resorb between 40 and 53 gallons (150–200 l) of water every day. Aquaporins also help cells regulate their volume and pressure.

Agre shared the 2003 prize with Roderick MacKinnon (1956–), professor of neurobiology and biophysics at Rockefeller University, who has done groundbreaking work on the structure of ion channels. The chemistry of membrane proteins makes it difficult or impossible to get them to form crystals, which means that the main methods used to investigate protein structures cannot be used. In 1998, however, MacKinnon managed to obtain a high-resolution image of the structure of an ion channel. This solved a mystery about the molecule's behavior: how it could allow potassium ions to pass through while blocking the entry of sodium ions, which were smaller. He discovered that the inside of the channel binds to potassium atoms in a way that mimics how they normally bind to water molecules. One way to think of this is to imagine trying to fit a metal ball into a hole. If the hole is too small, it cannot be forced through, but if the hole were lined with cogs and the ball had a coglike surface, the teeth could mesh and allow the ball to pass through. The atoms that allow potassium to pass through the channel function in a similar way.

A very important job of ion channels is to allow nerve cells to communicate with each other. Another is to coordinate the activity of muscles, such as the beating of the heart, and the release of hormones into the bloodstream. Mutations in channel molecules cause diseases such as cystic fibrosis, in which a defective protein called CFTCR prevents cells from properly controlling the passage of chloride ions. This leads to problems in the production of sweat, mucus, and digestive juices that eventually cause fatal lung infections at an early age. The disease is recessive, which means that a person usually needs two defective copies of the gene (one inherited from each parent) to develop symptoms.

The Discovery of the Water Channel

Many of the great discoveries of science have been made by researchers who were looking for one thing and happened across something completely different. As a young scientist, Peter Agre became interested in blood cells and the proteins on their membranes. This work led to the discovery of water channels, which he had not been looking for.

These channels had eluded researchers for many years. In the early 1950s the British biophysicists Alan Hodgkin (1914–98) and Andrew Huxley (1917–), working at the University of Cambridge, were studying the long axons of nerve cells in giant squids. They thought that the cells made an ideal system in which to understand how electrical impulses could be generated and transmitted long distances along the membrane of the nerve. They reasoned that charged atoms of sodium and potassium had to be passing in and out of the cell. These particles could not be simply slipping through the membrane—its chemistry and structure prevented that—and besides, if they moved freely, the cell would be unable to develop a strong charge. Hodgkin and Huxley's discoveries about the role of ions in the nervous system led to the 1963 Nobel Prize in physiology or medicine, shared with the Australian John Eccles (1903–97) for his work on the same theme.

The pioneering work on nerves led to the discovery of a wide range of ion pumps and channels. At first glance, water molecules seemed to require a different mechanism for entering cells. Often they could simply pass through the membrane, but some types of cells, such as red blood

(continues)

(continued)

cells, accumulated extra water and swelled. Researchers suspected that this was due to a channel protein, but no one could find it.

In the early 1980s Agre, a young doctor working at Johns Hopkins University in Baltimore, was working on membrane proteins in red blood cells. Agre, a biking enthusiast with a young family, had decided to go into molecular research rather than becoming a physician, and supplemented his income by working as a ringside physician at boxing matches. Agre became interested in the *Rhesus* (Rh) *factor,* a protein found on the surface of the blood cells of some people but not others. The effects of this molecule had been known for a long time. If a person who did not have the factor received blood from someone who did, his or her immune system would reject the blood, causing a severe reaction that could be fatal. The blood of a fetus might bear the factor, but it might be missing in the mother, which could also cause dangerous immune reactions. Normally, the circulatory systems of mother and child are kept separate during pregnancy, but blood sometimes is exchanged during birth.

Peter Agre, discoverer of the water channel *(Johns Hopkins University)*

When Agre learned that the Rh protein had not yet been found, he began looking for it among the membrane proteins of red blood cells. His laboratory succeeded in isolating two new molecules, the larger of which was the Rh protein. Agre's lab worked extensively on Rh, but he did not forget the smaller molecule. The team discovered that blood cells contained an amazing number of copies of the protein—about 200,000 copies per cell—and Agre wondered if it might be a channel. It was somewhat surprising that no one had seen the molecule before, but this was owing to the fact that it did not react to the normal stains used in the laboratory and had thus remained invisible. The lab discovered that the molecule was also produced in high numbers in kidney cells. When Agre discussed the molecule with his former professor John Parker, of the University of North Carolina, Parker suggested it might be a water channel.

Agre immediately set up experiments to test the idea. Greg Preston, a postdoctoral fellow in the lab, cloned the molecule's gene and inserted it into the genome of egg cells from a frog, *Xenopus laevis*. These cells had been selected because very little water normally moved through their membranes—possibly a sign that the cells had few or no water channels. If the new molecule were truly a channel, adding it to the cells should change their behavior. When the altered eggs were dropped into distilled water, they immediately swelled and burst—a sure sign that Agre's lab had discovered the water channel, which they christened "aquaporin."

In 2003 Agre received the Nobel Prize in chemistry for his work on aquaporins. Despite having received the world's most prestigious scientific award, he remains a modest and humble man. Invited to give a major talk in Berlin in early 2008, he spoke a bit about his work on water channels and other scientific themes. As well as singing Tom Lehrer's song "The Elements," a musical version of

(continues)

(continued)

the periodic table, he talked a great deal about his family and colleagues, who he claims have played an enormous role in his success. Referring to the Nobel Prize, he said: "I was contacted by hundreds of scientific colleagues, relatives from the U.S. and Scandinavia, friends from childhood, classmates from grade school through medical school—many of whom I had not seen in years. For me, the chance to renew these bonds is perhaps the best part of winning a Nobel Prize."

NEUROTRANSMITTERS AND SYNAPSES

Membrane receptors and ion channels work together so that nerve cells can communicate with each other, permitting thought and conscious movement. The signals that pass between them involve both small molecules called "neurotransmitters" and electrical stimulation.

Neurons make up only about 10 percent of the cells in the brain, but they have traditionally been considered the most important communicators. They have an unusual, treelike structure that sprouts from a small cell body, the soma, which contains the nucleus. One long extension, an axon, is used to transmit signals, and a huge network of rootlike *dendrites* receives signals from the axons of other cells. These structures meet at a tiny gap called a "synapse."

Most communication between the cells happens via neurotransmitters, which are released by the axon of one cell, flow across the synapse, and bind to receptors on a neighboring dendrite. Information flows one way because dendrites are not normally able to release neurotransmitters. On the other hand, an axon can reabsorb neurotransmitters that it has released,

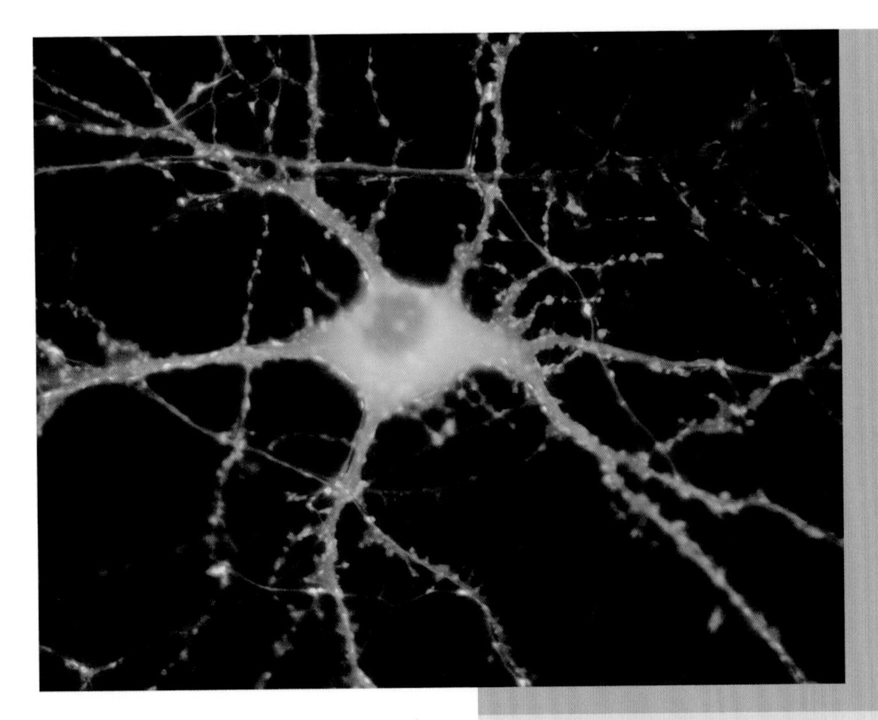

An image of a neuron from the University of California, San Diego, Division of Biological Sciences *(Michael A. Colicos, UCSD)*

which gives the cell a way to tune down the strength of a signal. Different types of cells—and different parts of the brain—have receptors for unique sets of neurotransmitters. This allows multitasking; it permits the brain to guide and keep track of different stimuli simultaneously. Some interactions are excitatory; in other words, they increase the electrical activity of the neighboring cell. Others are inhibitory, reducing the activity of the second neuron. Often information is passed between single neurons, but a special class of "modulatory neurons" broadcast neurotransmitters that spread more widely and can affect many cells.

When a neurotransmitter docks onto its receptor on another cell, it usually influences the behavior of an ion channel. Axons and dendrites are rich in these membrane proteins. High numbers of channels in some regions mean that parts of the membranes undergo very strong, rapid changes in electrical charge. This creates an impulse called an "action potential." It travels rapidly to the other end of the nerve, where it can be passed along to the next cell.

The action potential arrives at the axon side of the synapse and opens different types of ion channels, causing positively charged sodium and calcium ions to flood in through the membrane. Neurotransmitters are waiting just under the surface, packed in tiny bubbles made of fat molecules, lipids. The sudden change of charge draws the bubbles to the cell membrane, and they merge with it. This releases their contents into the synapse. Neurotransmitters drift across to the neighboring cell and dock onto their receptors. They open and close particular channels on the neighbor's membrane, changing its charge and causing the cycle to repeat itself in the new cell.

If a neuron is stimulated over and over again in the same way, the result may be long-term potentiation (LTP), which is like a kind of learning. The cell begins to amplify the signal. Usually this happens because of a feedback loop: Receiving more signals causes the cell to make more receptors for a particular neurotransmitter. This increases its sensitivity and creates a more powerful action potential. The opposite may happen to a nerve that does not receive a particular signal for a long time. It may make lower numbers of one of its receptors in a process called long-term depression (LTD). LTP and LTD are thought to play important roles in memory and learning.

The following is a partial list of neurotransmitters and some of their effects on the brain. The list has been simplified, because what actually happens when a neurotransmitter is released depends on the region of the brain involved and combinations of signals that are being received at any one time. It is sometimes difficult to distinguish between neurotransmitters and hormones, which are also small proteins. Additionally, most of these molecules are used in different ways by other types of cells throughout the body, so the list of known neurotransmitters is continually being updated.

- Acetylcholine is used as a signal to stimulate deliberate muscle movement.

- Dopamine is a signal often associated with feelings of pleasure, other moods, attention, and learning. It plays an important role in addictions.
- GABA (Gamma-aminobutyric acid), an amino acid, inhibits signals that control muscle movements.
- Glycine, another amino acid, serves as a neurotransmitter involved in reflexes and motor behavior.
- Glutamate, also an amino acid, is believed to play a role in memory and learning.
- Norepinephrine helps govern wakefulness and arousal.
- Serotonin is involved in memory and the sleep/wake cycle.

Until recently, researchers believed that neurotransmitters and their receivers were used almost exclusively by neurons. These cells are outnumbered about 10 to one by glial cells, which were formerly thought to have only structural and helper functions. This proportion is one source of the famous—and incorrect—idea that people only use about 10 percent of their brains. But recent research shows that the role of glial cells in brain activity has been significantly underestimated. Some of them release neurotransmitters such as glutamate. They also help remove neurotransmitters from the space between cells. Very recent studies suggest that they may even be capable of developing into new neurons.

Several diseases have been linked to defects in neurotransmitters or their receptors. For example, an area of the brain called the "substantia nigra" contains most of the neurons that release dopamine. These cells die in patients who suffer from Parkinson's disease, disrupting the dopamine system in the brain. Because the neurotransmitter helps control movement, people with the disease gradually lose control of their muscles. Amyotrophic lateral sclerosis (ALS), commonly known as Lou Gehrig's disease, develops when the body produces too much glutamate. People whose bodies produce too little serotonin often suffer from depression and other emotional disorders. Several hallucinogenic drugs, including LSD and mescaline, have their effects by binding to serotonin receptors and preventing

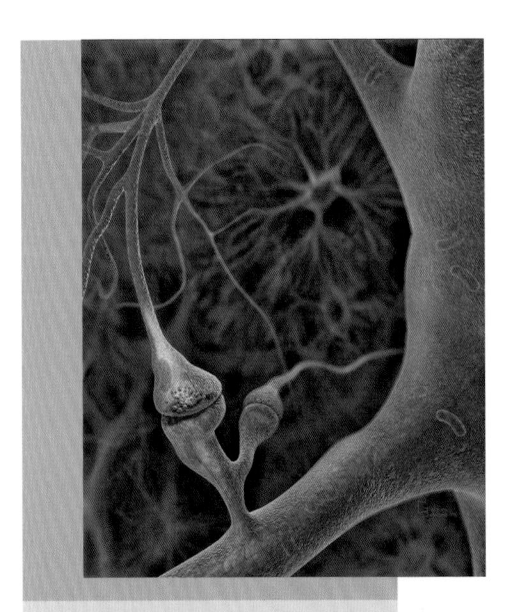

Deep inside the brain, a neuron prepares to transmit a signal to its target. Graham Johnson created this elegant drawing of that fleeting moment based on ultra-thin micrographs of sequential brain slices. After scanning a sketch into 3-D modeling software, he colored the image and added texture and glowing lighting reminiscent of a scanning electron micrograph. *(Graham Johnson)*

normal signals from reaching them.

A number of drugs and medications function by imitating the effects of natural neurotransmitters, blocking their receptors or preventing cells from reabsorbing them. Selective serotonin reuptake inhibitors (SSRIs), such as Prozac, slow down or block the process by which axons reabsorb the neurotransmitter serotonin, which leaves it in the system longer and increases its activity.

Because the nervous system is crucial to an animal's activity and survival, many of the toxins that have naturally evolved in scorpions, bees, snakes, spiders, and other organisms function by blocking the activity of neurotransmitters or their receptors. This paralyzes the victim and makes it easier to capture. Tetrotodoxin, a substance produced by several poisonous fish, binds to the pores of sodium channels and blocks them. Batrachotoxin, derived from the golden poison frog of South America, is one of the most potent poisons known. An amount equal to about two grains of table salt entering the bloodstream suffices to kill an adult human. It functions by allowing high amounts of sodium ions to enter the cell. This releases huge amounts of the neurotransmitter acetylcholine, overcharging the nervous system with signals, leading to fatal cramps of muscles and the heart.

MOLECULAR AMPLIFIERS AND VISION

Humans perceive the world as a colorful, noisy place full of smells and other stimuli thanks in large part to specialized receptor proteins that transform light, sound, and scents into energy or electrical stimulation. (GFP and other fluorescent molecules, described in the first chapter, do the opposite by converting energy into photons.) Many of the proteins that receive sensory stimulation have similar structures and functions.

Vision in humans and mammals depends on rhodopsin, which belongs to a family of transmembrane proteins, the G-protein coupled receptors (GPCRs). These molecules have a common structure: a domain outside the cell, a region that crosses the membrane, and an intracellular tail. Their transmembrane domains are always built of seven alpha helices that stitch a serpent-like pattern in and out. An incoming signal changes the structure of the tail so that it triggers a set of cellular amplifiers called "G proteins." These molecules behave like the kinases—introduced earlier as signaling proteins that activate other molecules by loading phosphate groups onto proteins and stripping them off again. But kinases use ATP energy—the *A* stands for the fact that they bind to adenosine, which is made of the nucleotide adenine and a sugar. G proteins bind to a different nucleotide—guanine—and use energy called *guanosine triphosphate,* or GTP.

It usually takes a combination of three different types of GPCRs to capture and transmit a signal. In the "quiet" mode, the three molecules are docked onto each other in the membrane. When activated by a proton or another stimulus, the group splits up into the active form, with two of the receptors bound together and one alone. Now they activate a G protein, which serves as a signal amplifier. It can go on to activate up to 100 new target molecules. In the case of rhodopsin, this means that the signal from a single photon can trigger the neuron to transmit an impulse to the next cell.

People perceive a range of colors and shades because of differences in the photopigments and the structures of light-sensitive cells. The human retina contains about 120 million

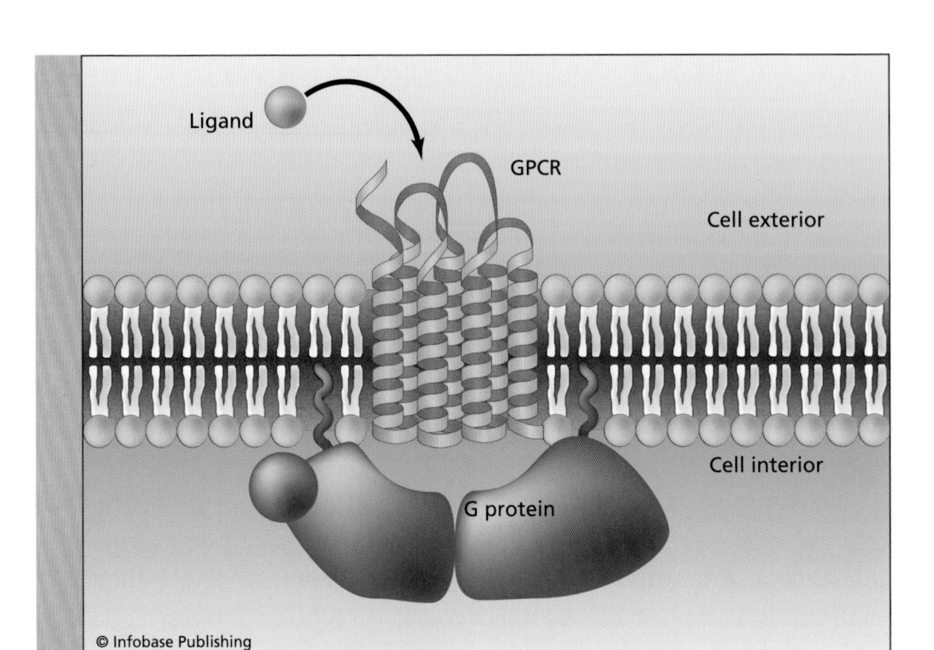

G-protein coupled receptors have a common structure: seven alpha helices crossing the cell membrane. They use G proteins to pass signals. Once activated, a G protein can go on to trigger up to 100 more molecules. This makes GPCRs very sensitive and allows them to detect tiny amounts of signal or small changes in a signal's intensity.

photoreceptor cells called "rods" and 6 million cone cells. Cone cells contain photopigments, small proteins that are stimulated by specific wavelengths of light. A single cone cell produces one of these pigments, allowing it to detect both a particular range of wavelengths of light and also the intensity of the light. Interpreting this information as color is up to the brain, which recognizes particular shades by mixing input from different cells. A rod cell is topped by a long, tube-shaped extension that looks a bit like a comb standing on end—the "teeth" are rows of membranes loaded with rhodopsin. Rods are much more sensitive, but they react to differences in the brightness rather than the wavelength of light. This makes them the main sources of stimulation in dim lighting conditions. The superb night vision of many species can be attributed to the high number of rods in their eyes. Some types of owls, for example, can see 100 times better than humans see at night.

Most people have genes for three color pigments, giving them the ability to detect differences in red, green, and blue (RGB) light. (RGB computer printers mix the same three colors to create pictures containing the entire spectrum of visible color.) But some people lack one or more types of cones, which makes them blind to the particular color range it senses. The genes for these molecules are located on the X chromosome, so men are far more likely to be color blind than women, because men have only one X chromosome. Women have two, which gives them two copies of the genes required to make different types of cone cells. If one is defective, the other gene can step in to provide color vision.

While some people lack certain cones, others have extra genes that permit them to see additional colors. This is well known in the animal world: Bees have a fourth photopigment that allows them to see frequencies of ultraviolet light invisible to humans; butterflies have five photopigments. Scientists suspected that extra pigments may have evolved through mutations in humans as well, and in 1993 Gabriele Jordan and John Mollon, two geneticists at the University of Cambridge, put the idea to the test. They predicted that an additional cone would allow people to distinguish between shades of colors that looked identical to most. In an experiment they asked subjects to tell whether two shades matched or not. One participant, known as "Mrs. M," seemed to be a "tetrachromat," with a fourth photopigment allowing her to see an additional color between red and green. There may be many more such people, but they may be hard to find in a world whose languages, art, and culture have been shaped by people with RGB vision.

People with the very rare genetic condition monochromatism are born without any cone cells at all. This means they cannot see colors; they detect only differences in brightness. Normally this problem affects only about one in every 30,000 people, but on the small Pacific island of Pingelap, nearly 10 percent of the population has inherited monochromatism. Scientists attribute this to a tsunami that washed over the island in the late 18th century, killing about 900 of the 1,000 people living there. Most of the survivors belonged to the royal family, in whom centuries of inbreeding had spread the mutation at a high rate. Many people living on the island today are their descendants. The story is recounted in Oliver Sacks's 1997 book *The Island of the Colorblind,* which contains fascinating anecdotes about the condition's effects on island art and culture.

GPCRs and photopigments are the basis of vision in mammals. Bacteria and many other unicellular organisms are also sensitive to light, although a different mechanism is responsible for stimulating the cell. Receptors on their surface bind to retinal, a small pigment molecule. When activated, it makes a swiveling movement that pumps protons—the positively charged subunits of atoms—from the inside of the cell outward, through the membrane. These are ion pumps, discussed earlier in the chapter, and they push protons into the cell even when the charge of the cell is already more positive than the surrounding environment. The effects are often like those of neurotransmitters. An action potential builds up in the membrane and travels down the cell to points of contact between neighboring cells.

Although the stimulus arrives in different ways, what happens next is similar in a wide range of species. The pathways that send light information to genes rely on the same basic set of proteins. This is strong evidence for evolution. Even more, it supports the hypothesis that the eyes of most—probably all—animals alive today originated in a single light-sensing structure in an ancient common ancestor. There is no reason that evolution could not have produced vision several times, but there is almost no chance that it would have used the same molecules to do so in very different species, unless they had inherited the genes from a common ancestor.

CHEMOTAXIS

The human genome encodes about 1,000 different types of GPCRs. In addition to transmitting sense information to the brain, some of the molecules act as "map readers." Different regions of the body are marked by specific molecules. Slight

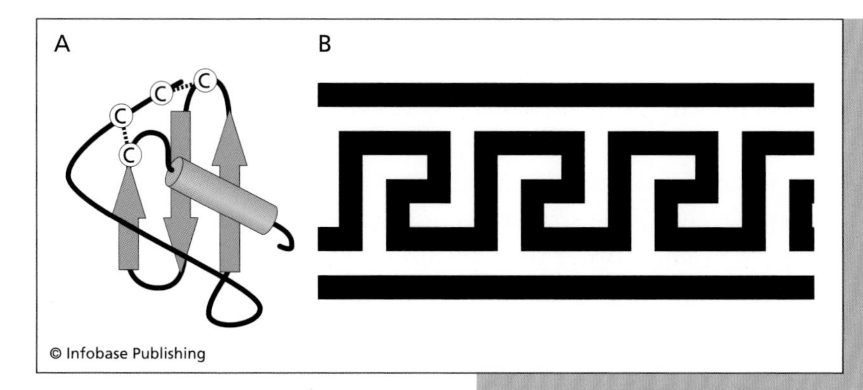

Humans have more than 40 types of chemokines, small signaling molecules that are used as a guidance system through the body. Chemokines have a typical "Greek key" structure (A), resembling a decorative pattern commonly used in ancient Greece (B).

changes in their intensity can be detected thanks to the signal-amplifying qualities of GPCRs. This influences cell migrations the same way that "sniffing" helps a person move toward the source of a smell. Migrations based on moving toward (or away from) higher concentrations of a molecule are called *chemotaxis.*

Chemotaxis is used throughout the development of the embryo and in some key body systems over a person's lifetime. One of the most obvious cases is the immune system, in which white blood cells learn to recognize and track down invading viruses and bacteria. Receptors on their surfaces recognize fragments of bacteria or diseased cells that need to be found and digested. Another type of receptor detects *chemokines,* small molecules secreted by other cells. Chemokine signals may be released at sites of infections, like an alarm that summons police to the location of a crime. They are also secreted by tissues such as lymph nodes, which act as training ground and meeting point for immune system cells. Cells need to enter the nodes but only gain access if they can receive chemotactic signals. The same system helps them find partner cells once they are inside. T and B cells (two types of white blood cells) have to meet up in lymph nodes to activate immune responses, and they are guided there by chemokine receptors.

More than 40 types of chemokines have been discovered in human cells. They usually begin as parts of larger proteins; after being cut out or whittled down, they are secreted from the cell. All of them are small molecules, and all contain a "Greek key" structure named after a decorative pattern used widely in ancient Greece. The structure consists of three beta sheets lined up next to each other and an alpha helix that lies cross-wise over the top. A few examples of chemokines and their functions are listed as follows:

- Monocyte chemoattractant protein-1 tells white blood cells called monocytes to leave the bloodstream and enter surrounding tissue. As this happens, the cells specialize into macrophages: or cells that destroy foreign particles by digesting them.
- Interleukin-8 prompts the migration of another type of blood cell, called a neutrophil, out of the bloodstream.
- C chemokines attract immature forms of T cells to the thymus, where they are trained to recognize the difference between the body's native molecules and foreign ones.
- Fractalkine is secreted by cells in the surface layer (endothelium) of the linings of blood and lymph vessels. Unlike other chemokines, it often remains attached to the cell by a long tether-like tail. As well as attracting immune system cells, fractalkine helps attach them to endothelial cells.

Chemotaxis is also known to play a role in cancer. When cancer cells leave a tumor in a process called metastasis, they do not settle just anywhere in the body. Specific types of cancer cells have preferred tissues to which they migrate. Sometimes the reason for their destinations is easy to figure out; cells from colon cancer tumors, for example, pass through the liver. They are too large to leave, so they get stuck and often spawn a new tumor in this organ. In other cases, tissues are invaded because they have chemokines that receptors on the cancer cells recognize. Researchers hope that by identifying the signals and receptors, they will learn to disrupt them and block the process of metastasis, which is the most frequent cause of death in many types of cancer.

DEVELOPMENTAL SIGNALS AND THE ORIGINS OF ANIMALS

If cells were unable to read their locations in the body or respond in different ways to the new signals they encounter as they grow and move, they would all remain the same. There would be no complex animals. Colonies might form, but they would be composed of a mass of undifferentiated tissue. Embryos arise from stem cells that divide, migrate, develop, and interact properly with their neighbors. All of these processes depend on receiving and properly interpreting signals. Starting in about the 1980s researchers began to study embryonic development as a process driven by genes, and they kept finding the same molecules over and over, filling similar functions in very different animals. For the most part, humans did not have unique, special pathways that built human beings—rather than, say, apes, or flies—from their cells. It turns out that a surprisingly small set of central signaling pathways drive the development of different tissues in a single body. One of those, the Wnt pathway, will serve here as an example of the role of signals in development and show how signals sometimes contribute to disease.

Wnt proteins are small molecules secreted by cells of worms, flies, fish, frogs, and mammals such as mice and humans. (The name is a fusion of the names of two other molecules—int and wingless—that had been studied in mammals and flies before researchers realized that they were related.) Some Wnts function as morphogens—molecules that stimulate cells to change their shapes and forms, differentiate into specialized types of cells, and develop into organs. Morphogens are released by cells in specific locations and spread through the growing body, docking onto receptors. Their effects depend on their intensity, that is, how much of the molecule cells sense. Close to the source, concentrations are high; farther away, there is less of the morphogen. As with neurotransmitters, the receiving cell can increase its sensitivity by adding extra receptors. Combinations of morphogens arriving from different directions act as a three-dimensional coordinate system that tells each cell its exact position in the body and give it developmental instructions. Morphogens produced on different edges of developing hands, for example, cause cells to build fingers of different lengths. They also create the differences between the palm and the back side of the hand.

The pathway that the Wnt signal activates is found in all animals, but not in their unicellular relatives—eucaryotic cells, such as yeast. This probably means that it arose at the same time as the first animals, and Wnt's important functions suggest that it may have made it possible for them to evolve.

Humans produce 19 kinds of Wnts, which arose through duplications and mutations of an ancestral gene. These signals can activate at least 100 genes, depending on what other signals are present and what else is going on in the cell. There are many receptors that can sense Wnts, and they activate various molecules to send the signal to genes. Interestingly, most of these signals pass through a common "gatekeeper," a central signaling protein in the cell called beta-catenin. Beta-catenin's activity is crucial to how most Wnt signals are handled. When the cell does not sense Wnt, beta-catenin is held captive by other molecules; any free-floating versions of the protein are destroyed. Beta-catenin has important functions in this captive state; it is sometimes woven into a system of protein clasps and fibers that tie neighboring cells to each other, but a Wnt signal releases it. This can cause the ties to dissolve and allow cells to migrate. That needs to happen as tissues grow and take on new forms in the embryo, but there are many circumstances in which it should not occur. If cells escape from a tumor, for example, they can travel to other parts of the body and spread the cancer. Not surprisingly, many types of cancer have been linked to defective Wnt signaling.

The signal has a second function. It prevents beta-catenin from being destroyed when it is released. This allows the molecule to move to the nucleus, where it binds to other proteins that change the pattern of active and quiet genes.

One effect of the signal is to help stem cells reproduce. The body contains many types of generic cells that differentiate into specialized types, starting with the most generic—embryonic stem cells, able to produce any tissue in the body—down through a series of increasingly specialized types. But not all cells should differentiate; the body keeps stocks of various stem cells on hand to make new blood, skin, and other tissues or to repair damage. Wnt signals are crucial in maintaining the body's store of stem cells that will become skin, blood, and heart muscle. This gives it an important role in regenerating tissues and healing wounds. While many damaged tissues are able to heal in human embryos and newborns, most tissues lose this ability with age. One reason may be that cells respond differently to Wnt later in life. If researchers could learn to control these signals, they might be able to restore the ability of some tissues to heal themselves.

Interfering with the behavior of beta-catenin in mice and other laboratory animals has given scientists a look at its role in various tissues. In the skin, if beta-catenin cannot be activated, scratch wounds do not heal and hair follicles fail to form. If beta-catenin is too active, the skin produces too many follicles, stem cells reproduce too often, and cancer may develop. Irma Thesleff, a Finn researcher who first trained as a dentist and then as a developmental biologist, has studied beta-catenin's role in the formation of teeth, which arise from the same basic type of tissue as skin. In 2006 her lab showed that if beta-catenin is overactive, the teeth of mice grow continuously. A single root produced as many as 40 teeth.

In the skin, Wnt signals play a role in the survival of stem cells, their differentiation, and the formation of patterns; defects lead to cancer. These functions are found over and over throughout the body. Wnt helps shape the stomach, colon, and intestines from a simple, tube-shaped structure. The signals are essential in the intestines; they are needed to protect a population of stem cells that differentiates to form the tissue. But for the stomach to form, the same signals must be repressed, and

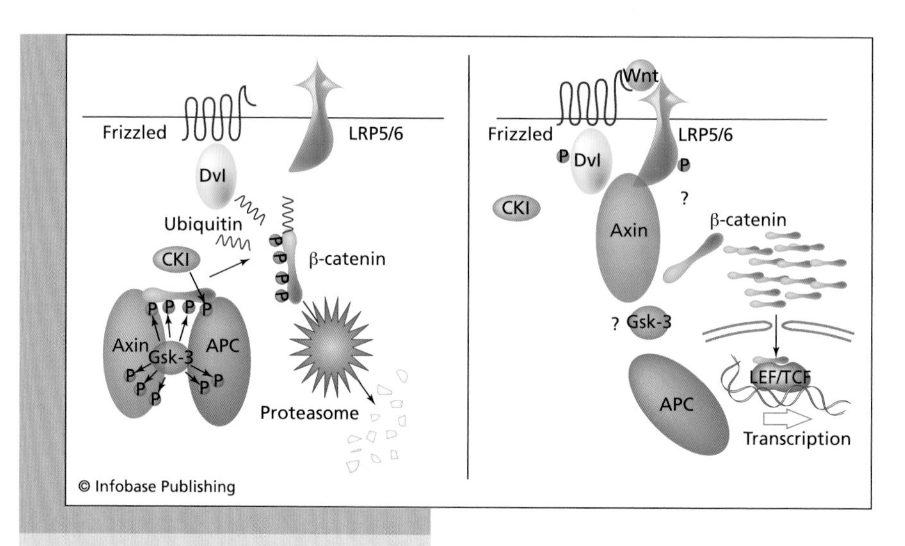

Small signaling molecules called "Wnts" play an important role in cell differentiation and the formation of animal bodies. Humans have 19 types of Wnt that pass signals to genes via many receptors and other signaling molecules. Most of these signals eventually reach the protein beta-catenin. In the absence of a signal (left), beta-catenin is held captive at the cell membrane or destroyed. When a Wnt signal is received, beta-catenin is released (right). It is protected from destruction as it enters the nucleus to activate genes.

defects in the system lead to intestinal and colon cancer.

Wnt and beta-catenin signals are equally important in the formation of the spinal cord. If beta-catenin is missing, the spine is too small; when it is too active, the spine becomes too large. Once again, Wnt's effects on stem cells seem to be the cause. Without the signal, the body cannot preserve its stock of stem cells; they differentiate too quickly and have been used up by the time they are needed. If Wnt is too active, there are too many stem cells, and the parts of the spine that hold them expand in size.

The pathway has similar effects on the development of structures in the brain, heart, bone, kidney, and blood. In each case, Wnt signals step in at a particular moment to preserve a certain type of stem cell or to push it to develop in a particular way. In turn, this prompts the body to build tissues and organs. Making these structures requires a careful control of growth, cell differentiation, and cell migrations, which is managed by signals. Losing control of them can lead to developmental defects, cancer, and other diseases.

Walter Birchmeier, a Swiss cell biologist who is director of the Max Delbrück Center for Molecular Medicine in Berlin, has studied Wnt signals for 25 years. "Wnt is one of a handful of powerful signaling pathways that have played a crucial role in the development of animal life and continue to widely control processes of embryonic development," he told the author in an interview. "It is not a 'master body-builder,' but it is an excellent example of the type of system that is needed to make animals. Nature uses a small set of tools over and over, in many parts of the body, in all types of animals. And it provides an excellent example of the connection between processes of normal development and disease. Switching Wnt signals on and off controls cell division, differentiation, and migrations in many tissues. These are exactly the processes that become defective in cancer. Our work is helping to show how some types of tumors are intimately connected to the deregulation of the pathway."

4

Traffic and Cell Architecture

One of the most beautiful and fascinating phenomena of nature is the mitotic spindle, a structure that cells build as they divide. Its function is to separate freshly copied chromosomes into two sets so that each daughter cell has a complete copy of the genome. In some types of cell division the spindle is built symmetrically around the center of the cell, drawing chromosomes outward in a mirror-image pattern that creates two identical copies of the parent. In other situations it forms off center, usually creating cells of different sizes. Perhaps the most amazing thing about the spindle is that for most of the cell cycle it does not exist. When it comes time for the cell to divide, other structures are taken apart, and their proteins are reconstructed into this shape. It is a perfect example of how the structure of a protein—the subunit used to build the spindle—permits a variety of things to be done with it and how a cell's form and structure are determined by the behavior of molecules. Interestingly, the same proteins that make up the spindle participate in other processes such as transporting molecules through the cell.

The mitotic spindle is not built by consulting a plan; it arises out of the self-organizing activity of proteins. The same is true of the other components of the cell's architecture. This chapter is devoted to the theme of how molecules work to give the cell its form and functions.

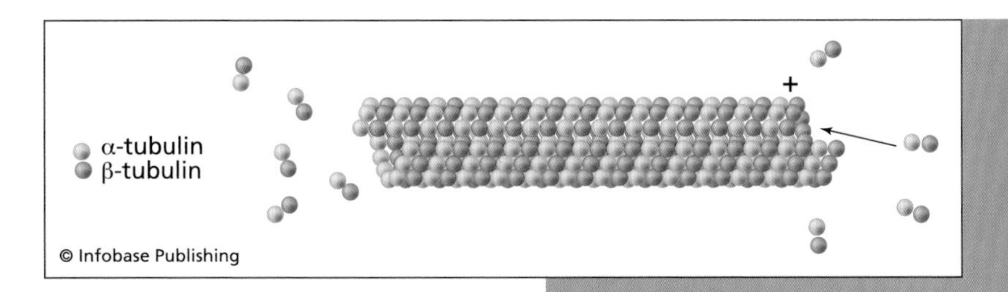

Mirotubules are an important structural element and traffickway in cells. They are built of single tubulin proteins, which come in two forms: alpha-tubulin and beta-tubulin. These proteins first form pairs and then assemble into tubes.

MICROTUBULES AND MOTORS

Different types of cells need unique shapes to do their jobs: A neuron, the basis of the brain's communication network, has a treelike structure that puts it into contact with hundreds or thousands of other cells. Red blood cells, which are basically bags full of the hemoglobin proteins needed to transport oxygen through the blood system, are highly elastic and doughnut shaped so that they can squeeze through tiny capillaries. These extreme shapes, and all the variations in between, are established by the *cytoskeleton,* an intricate network of fibers that spread through the cell. This section and the following ones present two of these systems, microtubules and *actin* fibers, and show how they guide the life of the cell.

Microtubules were briefly introduced in chapter 1 as cellular "railroad tracks" used as delivery routes through cells. These long structures are built of repeated subunits of a protein called *tubulin,* which is assembled into hollow rods a bit like the way bricks might be used to build a round tower. They can be thought of as a cellular subway, except that passengers travel along the outside rather than the inside of the tube.

Normally, microtubules are built outward from a structure near the nucleus called the microtubule organizing center (MTOC). Microtubules are continually being built and taken apart, like scaffolding that has to be moved to paint a huge building. Their length and precise shape at any particular time depend on the work of additional proteins that add or remove tubulin from the end or stabilize it to keep it from breaking down. But in preparation for cell division, the rules change. The entire network is disassembled. Microtubules are broken down into their subunits, which are then rebuilt to form the mitotic spindle.

Two types of tubulin protein (alpha and beta) are needed to build a microtubule. First, single proteins of each type are assembled into pairs. These are strung end to end in strands; the strands join up lengthwise to make a sheet, which then is bent into a tube. Microtubules can have different diameters: They can be made of between nine and 16 strands. The number can even vary within a single tube, giving it an uneven surface that narrows in some places and widens in others.

Each tubulin subunit can dock onto GTP, the energy-loaded molecule introduced in chapter 3. This activates it so that it can bind to a partner and then the growing microtubule. When that happens, GTP switches to its low-energy form, GDP. Mark Kirschner and Tim Mitchison, cell biologists at the Harvard Medical School, have shown that this transformation plays a crucial role in the stability of the microtubule. In three decades of studies of cellular architecture, they have discovered that microtubule subunits bound to GDP are normally instable and quickly fall apart. This can be prevented if the tip is protected by a caplike structure of tubulin subunits that still hold GTP. Kirschner and Mitchison believe that instability is caused by the fact that GDP-bound tubulins try to curl inward. This puts stress on the microtubule and may break it apart. Partner molecules can help stabilize the structure by binding to the cap. Other proteins destabilize it. Some drugs have their effects by binding to tubulin and disrupting the formation of microtubules, which influences the structure of the cell and processes such as cell division.

Many types of molecules can bind to individual tubulin proteins and/or the surfaces of microtubules. A number of these microtubule-associated proteins (MAPs) are required to trigger the assembly of the long tubes. These include the kinesins, a

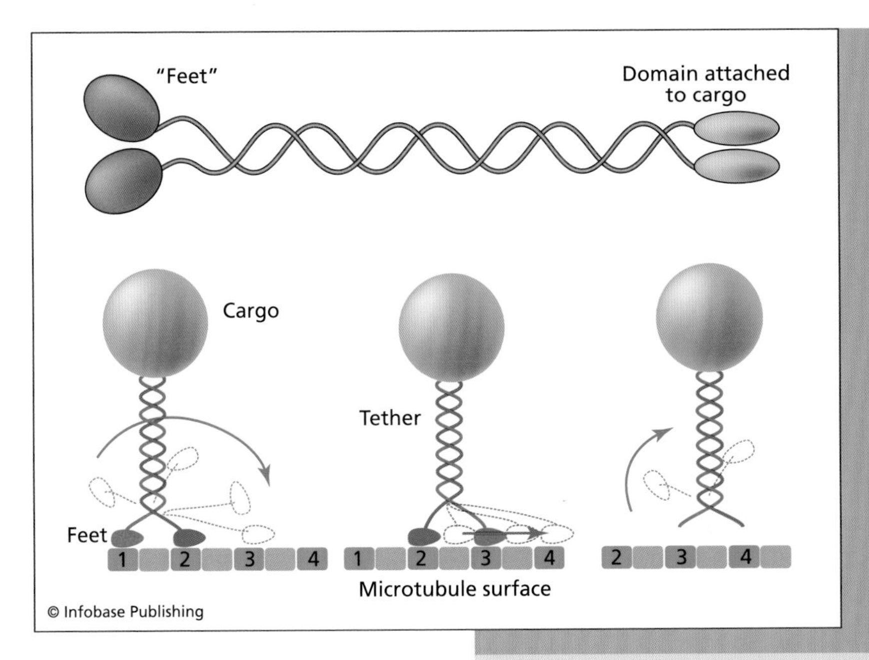

Motor proteins such as kinesin tow cargoes of proteins and other molecules through the cell by "walking" along microtubules. Two copies of the motor bind to each other to activate the molecule and give it two "feet." The motor moves by swiveling one foot over the other, then the second foot slides forward. The different steps avoid twisting the tether, which is attached to a cargo.

family of "motor" proteins that were briefly introduced in chapter 1.

A kinesin protein has a motor module that binds to the surface of the microtubule and a long tail (the towing line) to which cargoes bind. Building a motor requires two kinesins. Their two motor domains serve as "feet" that approach the surface of the microtubule and bind to it—pulling the cargo one step farther—and release it again. Before one foot lets go, the other has to bind. The free foot has to flip over the other, or swivel around it, taking another step forward and binding farther down the surface. Whether a foot binds or lets go depends on its shape and chemistry. The changes in structure that permit these two different states are driven by ATP energy, through the transfer of phosphate groups.

There are several types of kinesins as well as other protein motors. Each type regards microtubules as one-way streets; it can only travel in one direction. Most kinesins walk from the origins of the microtubules in the MTOC to the ends where new subunits are being added and taken away. Other motors, such as a molecule called dynein, move in the opposite direction. In 2003 Eric Karsenti, a cell biologist at the European Molecular Biology Laboratory in Heidelberg, showed that this feature of motors plays a key role in building the mitotic spindle. He wanted to understand how the physical characteristics of tubulin and the proteins they interacted with permitted this amazing architectural feat. His lab discovered that as a motor moves, its feet sometimes attach themselves to more than one microtubule. This draws the two fibers together and aligns them in the same direction. In a collaboration with Stan Leibler at Rockefeller University, Karsenti and his colleagues have modeled this behavior in the computer to show that complex cellular structures such as the spindle can be generated by a few molecules, following a few simple rules generated by their physical structure. In making the model, computer modeler François Nédélec and the rest of the team had to know the following:

- the probability that a motor will bind to a microtubule or detach
- concentrations of the molecules that are involved
- the rate at which a microtubule grows and the amount of flexibility it has
- the probability that a motor will fall off the end of a microtubule
- and the speed and direction at which the motor travels

Careful studies of cells gave the researchers the data they needed to establish some of these parameters, and their simulation of virtual mitotic spindles on the computer screen eerily mimics what they see under the microscope. The lab can now use the model as a way to test hypotheses about the behavior of microtubules and motors. Slightly changing one parameter on the screen—for example, slowing down a motor protein—may cause the spindle to break down and microtubules to assume a completely different shape. If the scientists notice similar structures arising in cells, they know where to begin looking for an explanation; for example, they can try to find another protein or something else that is slowing down the motor.

MICROTUBULES AND EVOLUTION'S TOOLBOX

Microtubules help give cells their shapes and functions in every plant and animal species. Until recently many scientists believed that this part of the cytoskeleton first arose with the development of multicellular organisms and other eukaryotes, because bacteria did not seem to have tubulin proteins. Until 1991 it was not even clear that they had a protein cytoskeleton, but that year Erfei Bi and Joe Lutkenhaus of the University of Kansas Medical School discovered proteins that help restructure bacteria as they divide. In the November 14, 1991, edition of *Nature,* they reported that a protein called FtsZ forms a ring-shaped structure around the "equator" of a bacterium as it prepares to divide. This beltway is drawn together and pinched off to create two daughter cells. A close look at the sequence of the FtsZ gene showed that it was so similar to tubulin that one molecule had likely evolved from the other. In 1998 Jan Löwe and Linda Amos of the Medical Research Council Laboratory of Molecular Biology in Cambridge determined the structure of FtsZ at exactly the same time as Eva Nogales's group at the University of California, Los Angeles, exposed that of tubulin. The shapes of the two molecules turned out to be remarkably similar, which means that they may well be related through evolution.

This raises a question: If "structural" proteins give cells their shapes (and they go on to sculpt tissues and organs), and tubulin has hardly changed, how could evolution produce animal bodies with such different shapes? Marc Kirschner and John Gerhart, professors at Harvard, have found an answer. They see microtubules as a prime example of how new forms and biological functions arise through the evolution of existing molecules and structures. In their 2005 book, *The Plausibility of Life,* they point out that tubulin itself has barely changed since the evolution of the first animals took place hundreds of millions of years ago, but species have evolved a range of unique, specialized molecules that regulate the growth and structure of microtubules in specific contexts. In other words, tiny changes can have huge effects.

A mutation in a regulatory molecule, for example, may influence how and where the spindle forms. Long ago such a mutation suddenly pushed the spindle to one side of the cell and caused it to divide in an asymmetric way; its daughters were no longer identical. That was crucial for the development of plant and animal life, because it is an important factor in the differentiation of cells into specialized types. Unicellular organisms are identical to their parents, but the cells of plants and animals undergo changes in shape and structure that can often be traced back to asymmetries in cell division. As Kirschner and Gerhart point out, the evolution of a new process does not require the cell suddenly to start making new molecules or undergo other very dramatic changes. It usually starts with a single change in an existing machine, which may have dramatic effects on the entire organism. The limbs of animals provide a good example. A common set of genes is used to build the pectoral fins of fish and the forelimbs of mice. Since that is the case, the authors ask: "Where do the differences arise in the limbs of bats, porpoises, horses, and humans? As one might expect . . . they come from the timing of, and amounts of, the secreted factors and selector genes affecting the growth of the various limb bones." By "selector genes" the authors mean molecules such as transcription factors that influence when, where, and how long particular genes are active.

The structure of the microtubule network is essential to nearly everything that happens in the cell. Mutations in the molecules that regulate them play such a vital role in so many processes that mutations in the molecules they work with have a powerful effect on cells and the bodies they build. Similarly, drugs that bind to them have very powerful effects. Cells can be brought to a standstill—and eventually destroyed—by making microtubules more stable or doing the opposite, causing them to break down faster than normal. Either event would prevent the cell from taking apart the network and rebuilding it into a spindle, or dismantling the spindle to construct transport routes through the cell. Scientists have discovered a wide range of substances that do this by binding to tubulin in different ways.

Colchicine, for example, is derived from the plant meadow saffron (*Colchicum autumnale*). It is a natural poison, but extracts of the plant have been used in small doses as pain relievers and as a treatment for gout since Roman times. Colchicine is still used in therapies for gout and some immune conditions. It disrupts microtubules by interfering with the first step in their assembly: the binding of alpha- and beta-tubulin molecules. The drug mainly binds to beta-tubulin, near the site where it docks onto alpha-tubulin, on the side of the sheet that will become the inside of the microtubule. By doing so, it changes the shape of the beta subunit so that new microtubules do not form. In high doses it even breaks up existing microtubules. It is not known whether colchicine binding pushes alpha and beta proteins apart or draws them close together; in either case, the change in shape causes the outer surface of the microtubule to become disorganized and accessible to enzymes that break down proteins.

Vinblastine was isolated from the sap of the Madagascar periwinkle flower (*Catharanthus roseus*). Doctors discovered that teas made from the substance reduced a person's number of white blood cells, suggesting that it might be useful in fighting blood cancer. Vinblastine seems to bind to the ends of filaments before they are linked into tubes. This prevents the subunits from carrying out the exchange of GTP to GDP, which is usually necessary for the addition of new filaments onto the growing tube. This probably changes the way new subunits are snapped on, as though trying to snap together LEGO building bricks that have melted and twisted. As a result, filaments curl into small spirals that cannot be linked together to make a well-organized microtubule.

Paclitaxel was extracted from bark of the Pacific yew tree (*Taxus brevifolia*), an endangered species, in 1967 as part of an initiative by the National Cancer Institute to isolate new

Why Bacteria Do Not Need Brains

Many species of bacteria resemble tiny fish, propelling themselves along in search of food with a tail-like *flagellum. Escherichia coli,* a bacterium that inhabits the human intestine, is covered with flagella that it uses to swim toward food and to escape from poisons. In animals such behavior would usually be coordinated by the brain, but bacteria manage without one.

Decades of studies by scientists such as geneticist Julius Adler (1930–), of the University of Wisconsin–Madison, have shown that bacteria manage this through the self-organizing behavior of proteins. "Swimming" in bacteria is controlled by signals and their effects on the structure and activity of bacterial molecules.

Receptor proteins on the surface of *E. coli* act as sense organs, tasting the environment for food—usually sugars or amino acids—or poisons. They also sense changes in the concentrations of these molecules through chemotaxis, described in the previous chapter. Adler called the proteins "chemoreceptors." They pass signals into the cell through methyl-accepting chemotaxis proteins, or MCPs. They have this name because the region of the protein that lies just inside the cell membrane can be tagged with methyl, a group of atoms made of carbon and hydrogen. Methyl serves as the receptor's "memory." When molecules called "Che proteins" tag it with a lot of methyl, the receptor becomes increasingly sensitive. When other Che proteins remove the tags, the protein is harder to stimulate. This accounts for its ability to sense different concentrations of a signal at different times and places.

substances from plants and test their anticancer properties. After many years of development it was turned into a potent anticancer drug by the pharmaceutical company Bristol-Myers

Ultimately, the tags affect swimming behavior through a surprisingly simple system. Each flagellum is mounted on a rotorlike structure in the membrane. The rotor can turn clockwise or counterclockwise. Each filament has a helical, corkscrew-like shape. Just as turning a corkscrew one way drives it into a cork and turning it the other way pulls it out, the two directions have different effects on the flagella. When they rotate in a counterclockwise direction, they bundle up with each other, creating a larger tail that propels a bacterium in a straight line. But when the rotation is clockwise, the bundle disbands, and each filament points outward. This causes the bacterium to tumble in place.

Counterclockwise rotations are caused by a signal that there is food somewhere ahead or something poisonous to the rear. This leads to straight swimming, and how long it lasts depends on the power of the signal. Usually it goes on for a few minutes, and then the flagella reverse directions and cause the cell to tumble. This allows receptors to taste the waters and reacquire the signal. If there is no signal, the bacterium alternates periods of swimming and tumbling, tracing a random path.

Adler and his colleagues made many of these discoveries by studying bacteria with mutations that disturbed particular elements of the system. Some changes in chemoreceptors made bacteria unable to sense and transmit signals from the environment. Defects in MCP molecules did not allow Che proteins to methylate the molecule, for example, and sent bacteria swimming in straight lines for days on end, regardless of changes in the signal, rather than the normal period of a few minutes.

Squibb. Paclitaxel makes microtubules more stable than usual and harder to break down. It binds to the inner surface of microtubules near the site where tubulin proteins dock onto GTP. It seems to create unusually strong ties between separate microtubule filaments. These normally become destabilized when beta-tubulin exchanges its GTP for GDP, but when paclitaxel docks on, the structures become much harder to dismantle. Usually, filaments try to curl when beta-tubulin is bound to GDP. This creates stress that hastens the process of breakdown, unless the end is stabilized by a cap. Paclitaxel may prevent the curling and stabilize the entire structure.

ACTIN FILAMENTS

The second major architectural system of the cell consists of a set of fibers made of the protein actin. Like microtubules, these filaments form a dense network that has several functions: to give cells shape and structure, to help muscles contract, to serve as a second delivery route for molecules, and to help cells migrate. These jobs are so crucial that evolution has tolerated very few mutations in the actin gene. In spite of hundreds of millions of years of evolution that separate humans and yeast cells, the sequences of their actin proteins are about 95 percent identical, as compared to a much higher rate of mutations and change among other molecules.

Actin filaments are the ropes and wires of the cell. They bundle together to tie things up and help give cells shapes that are important to their functions. For example, they support tiny projections called stereocilia on the surfaces of cells in the ear. These tubelike structures stand upright and sway under the influence of vibrations made by different frequencies of sound. The movement jostles membrane proteins, which then open and close ion channels. This changes the charge of the membrane and creates an electrical impulse that is passed from neuron to neuron until it reaches the hearing centers of the brain.

Another job of the filaments is to help cells bind to each other. Just inside the cell membrane, they act as an anchor for the molecule cadherin. The protein's tail is attached to the filament, and then cadherin passes outward through the membrane. Outside the cell the protein has a domain that acts as a tiny hook, snagging the hook of a cadherin on a neighboring cell.

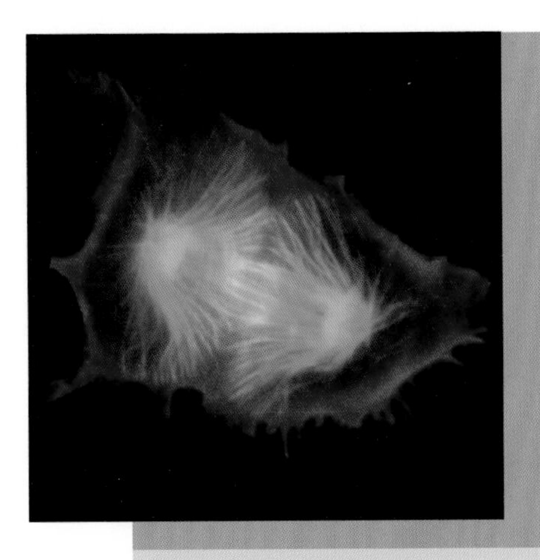

This cell has been triple stained for F-actin (red), microtubules (green), and DNA (blue). The cell was fixed about five minutes after the onset of anaphase, a stage in the cell cycle when chromosomes separate. *(Julie C. Canman)*

Actin fibers are known as F-actin, and they are built of single protein subunits called G-actin. The structure of G-actin was determined in 1990 by a group of researchers in Ken Holmes's laboratory at the Max Planck Institute for Medical Research in Heidelberg. Their map of the molecule revealed that the proteins likely twist around each other in a helical shape, like two vines, to form fibers.

Like microtubules, actin fibers are in a constant state of being taken apart and put back together. This behavior is controlled by a complex network of proteins that have been uncovered, in part, through studies of the *Listeria* bacterium. When it invades the cell, this organism uses actin filaments to its own advantage. It strings G-actin into long tails that push it through the cell, a bit like reaching for the ceiling by stacking one box on top of another, climbing up, reaching down for another box, and stacking and climbing on top of it. Interfering with molecules in infected cells has allowed scientists to pin down several of the molecules that switch on and off the construction of the fibers.

One of actin's jobs is to help muscle cells contract and relax. Skeletal muscle, which is responsible for voluntary movement, is built of long fibers that are created by the fusion of many muscle cells. The fiber contains millions of tiny, pistonlike subunits called

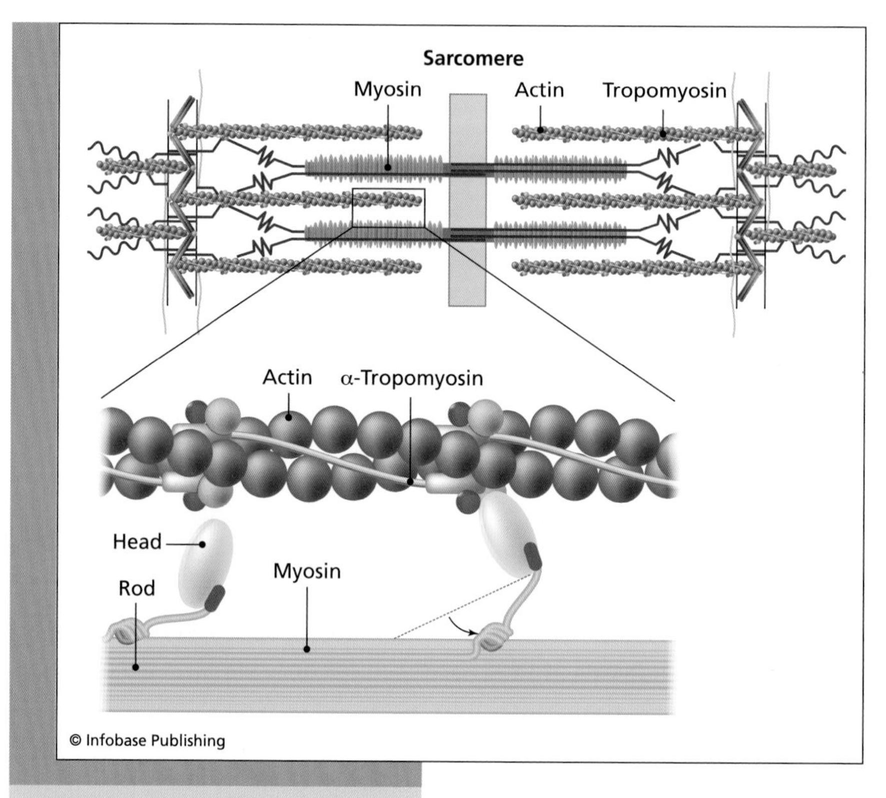

Actin fibers, or F-actin, one of the major components of the cell cytoskeleton, are assembled from protein subunits called G-actin. As well as binding cells together and helping them migrate, the fibers play a major role in the contraction of muscles. They help pistonlike structures called sarcomeres expand and contract. Actin does this by crawling along another type of fiber, made of the protein myosin, in a ratchetlike movement (described in detail in the text). The image at the bottom shows a close-up view of the interaction between actin and the "heads" of myosin proteins.

sarcomeres, seen in the accompanying image. At the core of each sarcomere is the piston's "rod," a thick filament made of a protein called "myosin." Around it, the sheath of the piston is composed of six actin filaments. Muscles contract and relax because the actin and myosin fibers slide along each other.

Individual myosin proteins look a bit like golf clubs. Their heads stick out of the fibers where they can bind to nearby actin filaments. The myosin heads can also bind to the energy molecule ATP, which drives their chemical activity. The sequence of events is as follows:

- A myosin head docks onto the actin filament.
- This changes the shape of the head and pushes the filament a small distance.
- The myosin head docks onto ATP, making it release the filament.
- ATP is converted to its ADP form.
- This restores the myosin head to its original shape and position. It binds to a new spot further along the actin filament, and the process starts over again.

Without ATP energy, myosin would not release the actin, and sarcomeres would freeze in place. This is what happens in rigor mortis, the stiffening of the body that occurs within a few hours of death. The supply of ATP to muscle tissue is cut off, locking myosin heads into place. The muscles only relax again when actin and myosin themselves begin to break down and lose contact with each other, after about 72 hours.

Voluntary muscles do not contract all the time because they have to be stimulated by signals from nerve cells called "motor neurons." When the brain sends a command (for example, to type a key on the computer or shoot a basketball), electrical impulses travel down these neurons until they reach the points of contact with the necessary muscles. The nerve cell releases the neurotransmitter acetylcholine, which crosses synapses to neighboring cells. This causes ion channels to open in the membrane of the muscle fiber, creating an action potential that spreads to other channels. After many more steps involving other molecules, this causes positively charged calcium ions to enter the sarcomere. The result is a thousand-fold increase in the concentration of calcium in the cell, which allows the interaction of actin and myosin molecules and thus the contraction of muscles.

Actin filaments link up to troponin, a protein that can bind to calcium ions. When this happens, the actin fiber twists, exposing the regions that are needed to dock to the myosin heads. Without a signal, the twist does not take place, and the actin and myosin do not bind.

Several well-known muscle diseases are the result of problems with the structure and function of sarcomeres. One of these is Duchenne muscular dystrophy, a genetic disease that occurs in about one in 3,000 males. People with the disease have a mutation in a gene called dystrophin, located on the X chromosome, which encodes a protein needed to properly connect the actin fibers to other parts of the sarcomere. People who lack a working copy of dystrophin suffer from a progressive degeneration in their muscles that usually becomes severe by their teen years.

Actin also has a role in several diseases caused by viruses. As copies of the vaccinia (cowpox) virus approach the cell, it reaches out with long, fingerlike protrusions that form because actin fibers push at the membrane. Proteins on the surface of the virus likely interact with receptors on the cell and activate signals that cause the fibers to form.

Like microtubules, actin fibers play an important role in the development of many specialized cell types. Actin forms the structural support for microvilli, small tubular extensions that increase the surface area of cells. This allows cells in the gut to absorb more food, and in the eyes it increases the sensitivity of photoreceptors to light.

Often the actin and microtubule systems have to work together in a coordinated way for cells to develop properly. In the late 1990s Carlos Dotti, a cell biologist at the European Molecular Biology Laboratory in Heidelberg, came across this situation as he investigated the structure of nerve cells. He wanted to know why neurons sprout only one axon but a large number of dendrites. This was curious because early in the development of the cells, the structures are identical: they begin as neurites, small, spike-like protrusions on the cell surface.

The answer lay in the cytoskeleton. An axon usually grows to be very long and has to be supported by microtubules, so it begins to grow when microtubules push at one spot on the cell membrane and elongate. This only happens in one place because normally a dense "thicket" of actin fibers prevents microtubules from reaching the membrane. In developing nerve cells, signals cause actin to break down at one place and microtubules make it through. By using drugs to break down actin artificially in other places, Ph.D. student Frank Bradke and other members of Dotti's lab could influence where the axon would form. They could also create cells with multiple axons.

"The stability of actin filaments at the growing edge of dendrites seems to be necessary to keep them from forming axons," Bradke told the author in an interview. "If you change that stability—as we did using drugs—a dendrite can be reprogrammed to become an axon." Many other types of cells, particularly those that migrate, have to coordinate the actin and microtubule systems in order to take on the right shape and move in the proper direction.

TRAFFIC INTO AND OUT OF THE CELL NUCLEUS

A membrane protects the cell from the outer world, holding together the molecules of life so they do not just float away and barring the entry of molecules that could be harmful. The membrane that surrounds the cell nucleus has a similar function. It divides the cell into two major compartments with unique collections of molecules and chemical properties. But molecules have to pass through all the time, at a rate of millions per minute. Messenger RNAs have to be moved to the cytoplasm for translation into proteins; transcription factors and other molecules have to be brought into the nucleus to alter gene activity. Most of this traffic passes through intricate channels called *nuclear pores,* which act as gatekeepers, checking the identity of molecules that move in both directions, allowing passage to some and barring the entry of others.

A nuclear pore is a basket-shaped structure whose sides are formed by eight proteins. They work with more than 20 other types of molecules in the nuclear pore complex (NPC). Each of the eight proteins extends a tail-like filament outward into the cytoplasm, where it interacts with other molecules; this process determines which molecules can enter and which are turned away. The transmembrane regions of the eight proteins link together and make a gap that small molecules can pass through easily. Large molecules usually need to be escorted through by helper molecules. In some

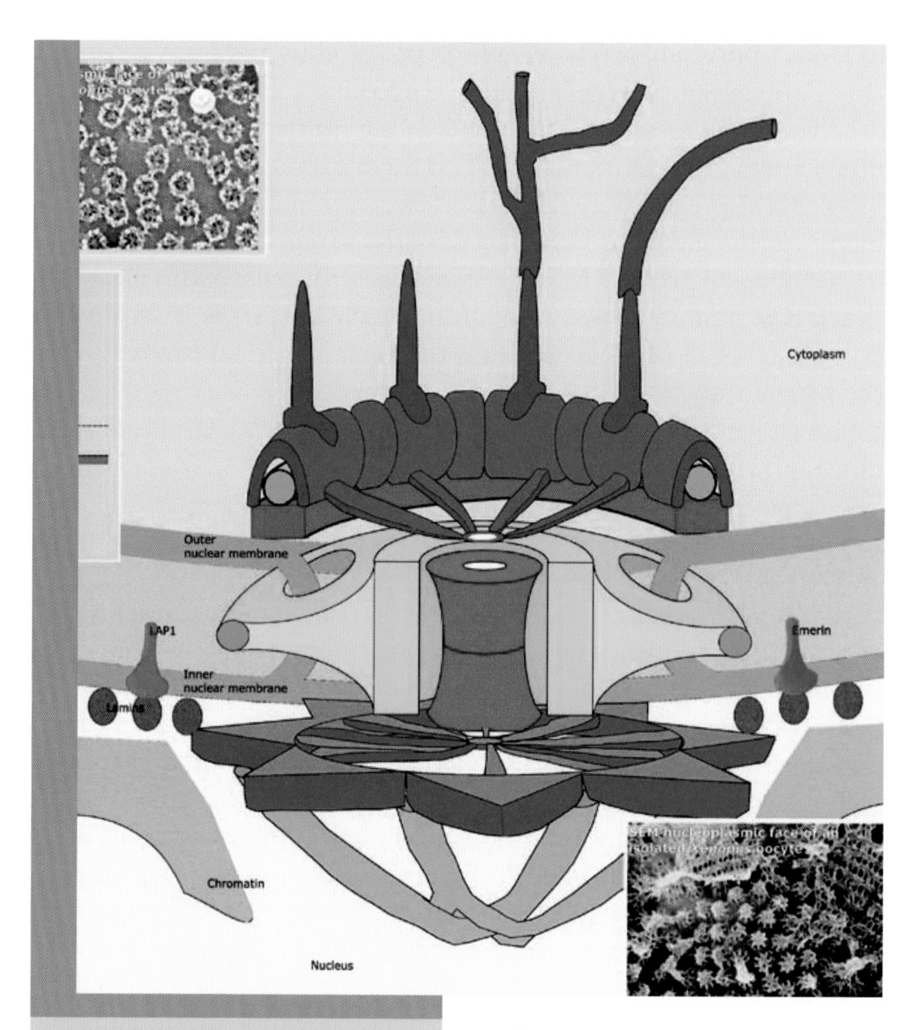

Nuclear pores are complex assemblies of many proteins that play an important role in the transport of molecules to and from the nucleus. By interacting with other molecules such as import and export factors, the pore proteins help decide what enters the nucleus, protecting the cell's DNA. This activity also prevents some defective RNAs from leaving the nucleus and being translated into proteins. *(Journal of Cell Science [2000], 113)*

electron microscope images the passageway appears to be plugged. This may happen because of changes in the shape of the pore proteins, or the plug might represent a cargo in transit. On the inner side of the nuclear membrane, NPC proteins extend fibers that draw together and form a ring, making a shape like a basket.

One important difference between the chemistry of the nucleus and the cytoplasm has to do with the activity of a protein called Ran. This molecule comes in two forms: a high-energy mode in which it is bound to GTP and a low-energy form bound to GDP. (This energy system was described in chapter 3 in the section "Molecular Amplifiers and Vision.")

Ran's energy mode plays an important role in the shuttle of molecules in and out of the nucleus. Proteins in the cytoplasm that need to be shipped to the nucleus contain a code called a nuclear localization sequence (NLS). The code is recognized by proteins called "importins," which bind to it. Together, they are now admitted by the NPC. Once inside the nucleus, Ran-GTP and a protein called CAS bump the importins off the cargo. This leaves the protein free to go about its business. It also leaves the importins bound to GTP, and they are recycled to the cytoplasm. Outside, the Ran-GTP is converted into Ran-GDP, which makes it release the proteins. They are now free to pick up new cargoes and start the cycle all over again.

When proteins have to be sent out of the nucleus, the process is similar. Outward-bound molecules contain a nuclear export sequence (NES) that attracts a carrier called an "exportin," which is bound to Ran-GTP. The exportin helps the molecule navigate through the nuclear pore. The same system is used to export most RNAs; they leave the nucleus with the help of proteins that contain the NES and dock onto Ran-GTP. Outside, the conversion to GDP releases the cargo, and exportin and Ran return to the nucleus.

Nuclear pores are stable machines that also help keep the cell cycle organized. Cell division requires breaking down one old nucleus and creating two new ones. The process is a bit like moving a huge puzzle by splitting it into many fragments, but without taking all the individual pieces apart. During cell division, regions of the nuclear membrane containing major parts of the nuclear pore remain intact. Later the fragments are retrieved (still containing their pores) and assembled into the two new nuclei of the daughter cells.

LIPIDS, MEMBRANES, AND VESICLES

This chapter and the previous one have shown how important membranes are in the internal and external business of the cell. In the plasma membrane surrounding the cell, receptors collect signals that allow it to respond to external events and guide its development. Channels and pumps manage the passage of ions and regulate the cell's overall charge. The nuclear membrane protects the cell's DNA and critically examines molecules that attempt to enter or leave. The cell has a wide variety of other internal membranes that manage its business, and some of them will be introduced here. But first, a word about how membranes are made.

The major components of the double-layered membranes that surround the cell and the nucleus are proteins, lipids, and cholesterol. Lipids are made in the endoplasmic reticulum by enzymes. They have hydrophilic heads (regions that can easily interact with water) and hydrophobic tails (that repel water); this means that they automatically twist and turn to try to put the heads in contact with water environments and to keep the tails away. Since both the insides of cells and the outside are predominantly water, lipids form a sandwich-like shape with the heads facing outward on both sides and the tails protected in between. This effect can be seen when one pours cooking oil into water: Its molecules quickly form beadlike drops because the fat molecules arrange themselves to point their hydrophilic heads outward and the tails toward the inside of the drop.

Lipids form double layers around the cell and the nucleus; they also enclose organelles such as mitochondria and the Golgi complex (a set of saclike compartments in which proteins are processed). Additionally, the cell creates a variety of other small, temporary compartments (vesicles) that are used to pack and transport molecules. An organelle called a *lysosome* is a sort of mixture between the two and is a good example of how membranes function inside the cell.

Lysosomes are membrane-enclosed compartments that are "pinched off" from the membranes of the Golgi complex, the way a person can pinch and twist a balloon to create a shape, for example, an animal. In the cell, vesicles close off completely and separate to become independent compartments. As this happens, they are packed with molecules that determine their functions. Lysosomes, for example, are loaded with dozens of powerful enzymes that can break down foreign substances or even old parts of the cell that need to be recycled. The cell uses these compartments to get rid of old organelles, invading bacteria, and other foreign substances. Some of their components can be used, so lysosomes are a source of nutrients and raw materials. For example, the cell obtains most of its cholesterol by removing it from clusters of proteins and fats floating through the bloodstream; they are drawn into cells and dissolved in lysosomes.

Because the enzymes in lysosomes can break down parts of the cell itself, under normal circumstances it would be dangerous for them to escape. On the other hand, they can be called on if the cell needs to be destroyed. *Apoptosis* (programmed cell suicide) is an essential part of animal development. It is the process that, for example, removes webbed tissue between a human embryo's fingers and toes before birth, and it helps completely reshape a tadpole's body as it becomes a frog. Enzymes from lysosomes help carry out these processes.

Normally they are held inside by the lysosome membrane, but there is a safety mechanism to prevent them from doing much harm if they escape at the wrong time. The interior of a lysosome is more acidic than surrounding parts of the cell, and its enzymes work best in this environment. The pH of the surrounding cytoplasm is usually about neutral, so if they leaked out they might not cause much damage.

Lysosomes are semipermanent structures; the cell produces many smaller membrane-enclosed packages for temporary use. Some of these vesicles make deliveries to lysosomes, bringing things that need to be broken down. The cell may trap a bacterium, for example, by engulfing it at the plasma membrane, in much the way neurotransmitters are taken up by nerve cells. A pocket forms on the cell surface. It deepens until it is pinched off, trapping the bacterium in a bubble called an *endosome*. (This is the term for inward-moving membrane compartments;

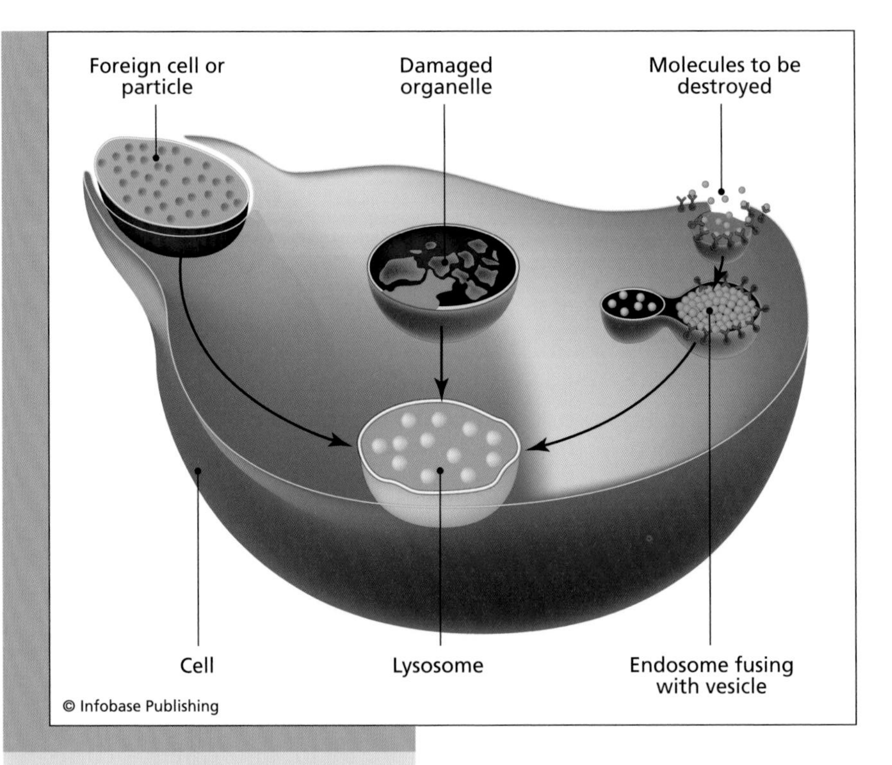

Lysosomes are cellular compartments that contain powerful enzymes that can destroy foreign particles or cells, damaged cell structures, and molecules that need to be destroyed. These objects are brought to the lysosome in membrane-wrapped compartments called vesicles. Proteins called SNAREs on the surfaces of the membranes ensure that vesicles are delivered to the right places.

outward-bound vesicles are known as *exosomes.*) Once the endosome arrives, its membrane fuses with that of the lysosome. The contents are dumped inside and taken apart by enzymes.

At any one time, a cell contains a huge number of endosomes, exosomes, and other vesicles. Keeping track of them and delivering them to the right places depend on "postal code" proteins mounted in their membranes. The delivery system is used, for example, to load lysosomes with their enzymes. After translation in the endoplasmic reticulum, the enzymes are packed into vesicles bearing a molecule called LAMP-1 and other protein labels that guide them to the lysosome.

LAMP-1 and lysosomes are a focus of the laboratory of Robert Siliciano, a physician and AIDS researcher at the Johns Hopkins University School of Medicine. Siliciano and his colleagues have been trying to take advantage of LAMP-1 and lysosomes to develop new types of therapies against HIV, tumors, and other health threats. One of the problems in developing a vaccine against HIV is that the virus inserts its genetic material into the DNA of the host cell, where it can lie dormant for a long time—Siliciano thinks it can hide there for up to 70 years. This tricks the cell into treating the "foreign" molecules of HIV the way it handles its own RNAs and proteins, rather than the way it treats other invading molecules. In 1995 Jennifer Rowell and other members of Siliciano's laboratory carried out an experiment to try to change this.

Their work started with the fact that the immune system recognizes most invaders when fragments of their proteins (*antigens*) appear on the surface of antigen-presenting cells (APCs). They are recognized by receptor proteins on white blood cells called T cells, which then orchestrate an immune response. (This process is discussed in more detail in the next chapter.) Before the APCs can "present" foreign molecules, they have to digest them, which involves bringing them into the cell in endosomes, delivering them to lysosomes, and breaking them apart. The fragments are combined with another protein called an MHC II (for "major histocompatibility complex"). But HIV proteins are usually not treated this way, since the cell makes them and does not recognize them as foreign. Rowell and her colleagues wondered what would happen if they could force HIV molecules to enter lysosomes, so they tagged one of the viral proteins with a sequence that would attach it to LAMP-1. The molecule was delivered to lysosomes and digested, then attached to MHC II and delivered to the membrane. The surfaces of the APCs now contained high amounts of the HIV protein—much higher than cells with the normal version of HIV. The same thing would probably happen in the body if researchers could somehow find a way to push parts of viruses into lysosomes. APCs coated with fragments of HIV proteins would probably evoke a strong immune response.

A similar strategy might work with cancer cells—which probably evade the immune system because they are produced by the body itself—and in the creation of new types of vaccines against other viruses or microbes. The laboratory of T. C. Wu, also at Johns Hopkins, has been trying to train immune system cells to recognize the DNA of foreign viruses by adding a code that delivers molecules to the lysosome. The researchers began with a gene called E7 from the human papilloma virus, which causes warts, cervical cancer, and other problems. They added sequences that would deliver the E7 protein first to the endoplasmic reticulum and then to lysosomes. Then they fired the DNA into APCs using a "gene gun." The artificial gene provoked a much stronger immune response than the normal E7 protein.

Several human diseases are caused by defects in lysosomes or the system that delivers molecules to them. Most problems arise because a person has a defective form of one of the lysosome enzymes, meaning that some substances cannot be broken down. Huge amounts of such molecules jam up the lysosome, which swells and stops functioning.

This is the problem in Hurler's disease, which causes bone deformities. Patients lack a protein called alpha-L-iduronidase. Examination under the microscope reveals that their cells are full of large compartments containing complex carbohydrate molecules that cannot be digested. Other missing enzymes result in a massive accumulation of lipids, animal starch, or other molecules that cause a wide range of diseases.

This section has dealt almost exclusively with lysosomes and endocytosis. Exocytosis—the outward movement of vesicles—is equally complex and has the same basic steps. Molecules that need to be secreted move from the endoplasmic reticulum to the Golgi complex and sometimes back again. Sometimes they make stops along the way; they may be unloaded to undergo processing and repacked again. "Codes" in proteins keep things moving in the right direction and tell the cell where vesicles are to be delivered.

Endocytosis and exocytosis have many functions, including the recycling of receptors and other molecules. While many receptors stay in the plasma membrane, where they can be activated over and over again, others are brought into the cell after they have been stimulated. They may have to be drawn in to meet up with a partner and pass a signal, or they may need to be "reset." If so, they are wrapped up in the membrane they are mounted in, drawn into the cell to another compartment, and then sent back once they have been "reloaded."

If all membranes that bumped into each other fused, the cell would develop a few massive compartments in which everything was mixed, and molecules would be lost all the time. Endocytosis and exocytosis only work because vesicles move in the right direction and deliver their cargoes to the proper targets. Once a vesicle arrives at its destination, the two membranes have to fuse. Both accurate delivery and fusion are managed in part by proteins called SNAREs. Human cells produce dozens of types of these small proteins. Packages to be delivered are labeled with vesicle SNAREs (v-SNAREs), and their destinations contain matching target SNAREs (t-SNAREs). When the two proteins recognize each other, they intertwine. SNAREs pass all the way through the membrane, and when they bind to each other, they create a zipperlike structure. They intermesh and open the membranes, which merge and bring the vesicle's contents inside.

Defects in SNAREs lead to disease. The bacteria that cause the deadly human diseases botulism and tetanus, *Clostridia,* carry SNARE-blocking toxins. SNARES are essential to the function of neurotransmitters, which are stored in vesicles until they need to be delivered to the cell membrane and released. By interfering with their transport, the toxins interrupt communications between cells, which eventually leads to death.

5

Molecules of Immunity, Health, and Disease

People usually notice a disease with the onset of symptoms such as fever, fatigue, or a problem with the heart, stomach, or another organ. But diseases begin as events involving molecules and cells, and by the time a victim knows that something is wrong, millions or billions of cells have usually been affected. This chapter introduces the new view of disease that has arisen in the molecular age and some new ideas about finding cures.

A person's health problems can be caused by parasites or infectious agents; they may also be the result of "problems with the system." This latter group includes aging and everything that accompanies it, such as most kinds of cancer that develop beginning in middle age, neurodegenerative conditions such as Alzheimer's disease, and problems with the heart and cardiovascular system. The body has evolved sophisticated defenses against infectious diseases but has little protection against systemic conditions, so different approaches are usually needed to treat the two types of problems.

Parasites such as viruses and bacteria have been a strong force in evolution; they are engaged in what has often been called an "arms race" with the plants and animals they infect. Many disease organisms need a host to survive. Some parasites, such as the Epstein-Barr virus, infect many people who develop no symptoms at all, although in others the virus causes mononucleosis, and if the immune system is challenged by a second threat, such as malaria, Epstein-Barr can lead to cancer. Real evolutionary pressure comes when infectious agents cause problems for their host; any plant or animal with a mutation that helps it survive is more likely to pass along "resistance genes." Then, when an animal evolves a new strategy to defeat a disease, the infectious agent often evolves a way around it. There is nothing conscious or mystical about this process. Disease organisms that cannot handle the defenses are wiped out and disappear, sometimes jumping to a new species. If a disease managed to kill all of its hosts before they reproduce, the species that it attacked would become extinct. All parasites and hosts that are alive today have avoided this fate, usually by adapting to one another through evolution.

Today the greatest killers in the industrial world are diseases caused by defects in the body's own molecules. About 5 percent of the world population has inherited a defect in a single gene that causes suffering, early death, or seriously alters the development of the body. Many other diseases are likely to be caused by combinations of genes, but these are extremely difficult to detect. Evolution has not provided much protection against these problems because they usually strike organisms after the age of reproduction, when they have already passed along the genes that are responsible. Curing these conditions will probably require new methods that replace or repair defective molecules.

THE INNATE IMMUNE SYSTEM

Innate immunity (also known as nonspecific immunity) is a very basic system for dealing with infections that evolved long ago in the ancestor of vertebrates and insects. The adaptive immune system evolved later and is much more sophisticated; it can learn to recognize a threat, mount a response, and remember it later. Innate immunity, in contrast, offers either/or protection: It recognizes and destroys an invader, or it does not. Its main actors are specialized kinds of white blood cells including mast cells, natural killer cells, and macrophages.

An innate immune response begins when the body notices that something is wrong. The alarm is usually sounded by complement proteins; about 20 types of these molecules patrol the bloodstream, looking for a foreign object to dock onto. Such binding has several effects. Some complement proteins act as alarm signals that call up white blood cells, which engulf foreign microbes or diseased cells through endocytosis, described in the previous chapter. A pocket forms in the membrane, and it deepens until the foreign organism is entirely surrounded. The invader is carried to a lysosome, where it is taken apart by enzymes. Sometimes destruction is accomplished by a "respiratory burst": The microbe is exposed to a high-energy form of oxygen that makes an organism's enzymes hyperactive, tearing them apart. Other complement proteins can crack open the membranes of invaders. The cells rupture and die when their contents spill out and molecules from the outside flood the cell.

A different response is triggered by cells that have been infected by a virus. They release proteins called interferons that bind to nearby cells. This activates systems in the neighbor that make it harder for the virus to enter and reproduce.

An additional line of defense involves white blood cell called natural killer cells. They recognize some cells that have been infected by viruses or taken over by cancer. They manage this because the diseased cells do not carry enough of a surface protein called *major histocompatibility complex* (MHC). Natural killer cells and MHCs are also involved in adaptive immunity and are discussed in more detail below.

Innate immune defenses are enhanced in tissues such as the skin, which comes into contact with the environment, by mast cells that react to damage by releasing histamine proteins. These molecules cause tiny capillaries to swell and leak blood into the surrounding tissue, which is why the site of an injury often undergoes painful swelling. When white blood cells arrive on the scene, they release other signaling proteins called "cytokines." One effect is to tell the brain to raise the body's temperature—a fever. This has an important immune effect because many microbes can only survive in a narrow temperature range; they are usually attuned to normal body temperature. Very high fevers may need to be treated. Suppressing milder fevers might make a person more comfortable, but at the same time it could make the body more hospitable to an invader.

Bacteria and some other infectious agents have evolved methods of slipping by innate immune defenses. The tuberculosis bacterium is enclosed in a shell-like capsule that protects it from being split open by complement proteins. Other parasites enter cells quickly, before they are discovered, then hide and grow inside them. Plasmodium, the one-celled parasite that causes malaria, escapes by growing inside red blood cells. It can only do so in properly formed cells, so people who suffer from sickle-cell anemia, in which red blood cells are improperly built, are at a lower risk of developing malaria.

ADAPTIVE IMMUNITY

Many microbes that escape innate immune reactions use the bloodstream to reach the cells they infect, so it is logical that blood should play a major role in adaptive (or specific) immunity. This is the body's second major system to protect itself from disease, and it has the following main properties:

- It can respond to a huge range of threats that cannot be foreseen.
- It must avoid attacking the healthy cells of the body it is defending.
- Cells that successfully recognize foreign proteins or other molecules make billions of identical copies of themselves, clones, to mount a full-scale attack.
- It remembers most previous infections and prevents them from successfully returning.

The main actors in this system are white blood cells called B and T cells, which come in many subtypes, including the macrophages discussed in the previous section. The system as a whole is based on the ability of B and T cells to distinguish native from foreign molecules ("self" from "not-self"). The system has to be flexible enough to recognize new threats; failing to detect an invader often causes serious disease. But if the system errs in the other direction and misinterprets one of the body's own proteins as foreign, the result may be an autoimmune disease. An adaptive immune response goes through a start-up phase in which it learns to recognize a new threat, then a phase in which it actively and broadly attacks the invader and, finally, has to be turned off again.

The body's "second circulatory system," the lymphatic system, plays a crucial role in adaptive immunity. This network of vessels and nodes stretches through the entire body, recapturing fluids that build up in the spaces between cells. It also absorbs fats. Lymph's waste-collection role puts it in a good position to act as a surveillance system for disease. The cells that flow through its vessels eventually return to lymph nodes, bringing along information about foreign molecules. The nodes are meeting points between roaming scouts called dendritic cells and B and T cells.

These cells are trained to distinguish self from not-self as they develop. They all arise from stem cells in the bone marrow. B cells remain there to mature, while T cells migrate at an early stage to an organ called the thymus, near the heart. Both types contain receptor molecules (antibodies on B cells are one type) that can potentially recognize foreign molecules. The receptors are built in a completely random way, in a process of cutting and pasting genes that is described below. The randomness means that many of the receptors would recognize the body's own proteins—which would be dangerous—if the cells were allowed to circulate. But as B and T cells mature in the marrow and thymus, those that might pose a threat are eliminated. Most are discarded; only about 2 percent of the T cells that are made survive the screening process and will be released into the body.

The structure of a lymph node plays an important role in immunity. It has an outer compartment called a T zone, mostly inhabited by T cells, and an inner compartment called a B follicle. Both types of cells enter the node through the T zone. Here they meet up with dendrites, which constantly roam the body in search of invaders. When a dendrite meets one, it digests the

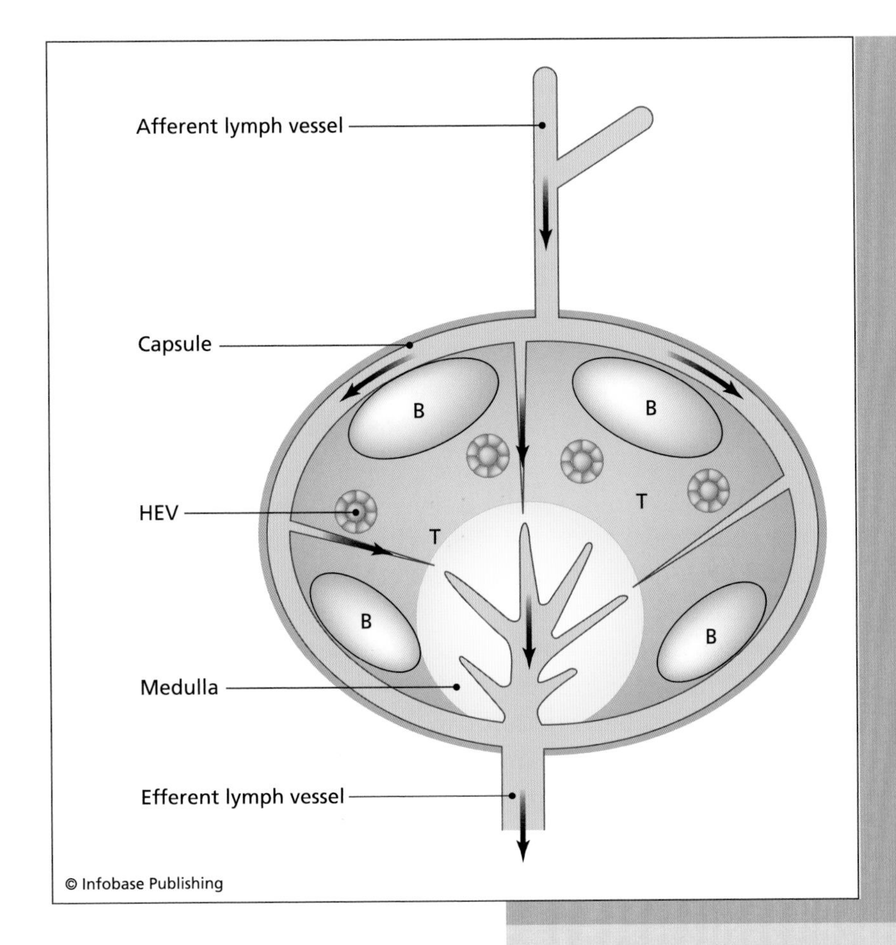

The lymph system collects fluids and wastes that have accumulated between cells and plays a crucial role in the animal immune system. Lymph nodes are meeting points for dendrites, which survey the body, and T and B cells, which seek out and destroy invaders. By passing through various zones in the nodes and encountering each other, immune system cells learn the difference between "self" and "not-self" so that they do not cause autoimmune diseases.

foreign molecules, chewing them into fragments and combining them with MHC molecules. This assembly is moved to the cell surface where it can be spotted by antibodies or receptors on other cells.

A full activation of the immune system usually begins at a lymph node, with the entry of a dendrite bearing foreign particles (antigens). If they are recognized by one of the randomly cut receptors on a T cell, the cell begins to divide into specialized types. Some of the daughter T cells immediately leave the node, enter the bloodstream, and set off in search of the offending microbe. Others, helper cells, move toward the border of the B follicle. If they meet a B cell whose antibody can recognize the same antigen, they activate it.

Activated B cells move into the follicle and also begin to reproduce. In doing so they make billions of new copies of the antibody. They leave the node and travel through the bloodstream, secreting antibodies along the way. These glue themselves onto the antigens on the surfaces of invading cells, pointing them out to white blood cells that will digest and destroy them.

Another group of T cells produced in the lymph node develops into regulatory T cells, which become active at the very end of an infection. Their job is to find and destroy other T cells left over from an infection. There are likely billions of these cells still around, actively seeking invaders. If they remain in the system, some of them are likely to undergo mutations or other changes that can easily lead to attacks against healthy cells. Autoimmune diseases can also be caused by problems with the thymus, the lymphatic system, or T and B cells.

Memory T cells are a specialized type that live for a long time and retain their ability to recognize an old infectious agent. If one enters the body a second time, the cells can divide immediately and mount a very fast attack, catching the organism before it does much damage. Vaccines make use of this system by causing the body to build memory cells before there has actually been a disease. They are usually made of dead viruses, weakened strains that are no longer infectious, or sometimes a harmless virus related to a dangerous one. When the vaccine is injected, the body is flooded by foreign proteins that are immediately noticed and cleared by immune cells. If the infectious virus then tries to gain a foothold in the body, the memory cells are activated, and it is quickly wiped out.

Researchers have not yet been able to make vaccines that effectively protect people against HIV, plasmodium (the malaria parasite), or many other infectious agents. HIV hides in—and eventually destroys—the very immune system cells needed to protect the body. Plasmodium goes through several stages of its life cycle during an infection, changing the proteins on its surface, like putting on a new disguise just as the body begins to recognize it. One experimental approach toward treating these diseases involves removing T cells from a patient's body and training them to recognize specific invaders, possibly even tumors. This strategy centers around outfitting the cells with new receptors to recognize a specific antigen and then reimplanting the cells in the body. The process is complicated for several reasons. First, the scientist must be sure to find a molecule that is only present on diseased cells. A second issue is to ensure that the new *T cell receptor* only recognizes the defective protein; otherwise, the cell that is supposed to be helpful may trigger an autoimmune disease.

HOW ANTIBODIES AND T CELL RECEPTORS ARE MADE

Chapter 2 introduced transposons, genes that can cut themselves out of an organism's DNA sequence, copy themselves, and reinsert themselves into new places in the genome. Cutting, pasting, and rearrangements of DNA also lie behind the production of antibodies and the receptors of T and B cells.

Antibodies are made by selecting bits and pieces of genes called immunoglobulins in B cells, a bit like assembling a ransom note by cutting letters from newspaper headlines. There are many immunoglobulin genes, and each cell recombines them in a random way, which explains how the body can make so many unique molecules. B cells produce receptor molecules that are made and function the same way; the difference is that receptors remain attached to the cell, whereas antibodies are released to float freely through the bloodstream. Each receptor and antibody is made of two identical halves that are slightly bent. Put together, they resemble a letter Y. The bottom half links the molecule to the cell membrane, and the upper half extends outward, where it will come in contact with molecules on other cells. Each half of the receptor contains two subunits. The heavy chain is longer, starting at the bottom, bending, and stretching up to form the two arms of the Y. Smaller light chains are attached to the arms.

When researchers first began to analyze the sequences of antibodies, they discovered that the molecules had very strange characteristics. First, they found that the cells of most vertebrates are able to make two basic types of light chains and five types of heavy chains. The different behavior of various types of white blood cells is partly due to the type of heavy chains they have (explained in more detail below).

It was no surprise to find that the body could make more than one type of antibody; the genome contains many duplicates of genes that have evolved slightly different characteristics. The curious thing was that each cell made a completely different molecule. The heavy chains have a lower section called the *constant region,* which comes in only five types, but every cell produced a unique version of its upper *variable region.* The same was true of light chains, which also had constant and variable regions. The variable regions make it possible for B cells to make nearly 20 billion unique antibodies.

Initially scientists were at a loss to explain how billions of different cells in the same person could produce billions of unique antibodies. Each could not be made from a unique gene; the human genome was not nearly large enough to encode so many molecules. In the 1960s and 1970s, laboratories across the world discovered how cells managed to create this huge number of individual molecules from a much smaller set of genes.

As scientists searched for the genes containing the constant and variable regions, they were surprised to find that the pieces were scattered at large distances from each other in chromosomes. Heavy chains are assembled from four different types of genes that begin as separate units on chromosome 14. The types are called variable (V), diversity (D), and joining (J) genes, plus the constant regions (C). As each B cell develops, the region of chromosome 14 containing these sequences undergoes a unique rearrangement. The process begins with the 12 D genes. One is

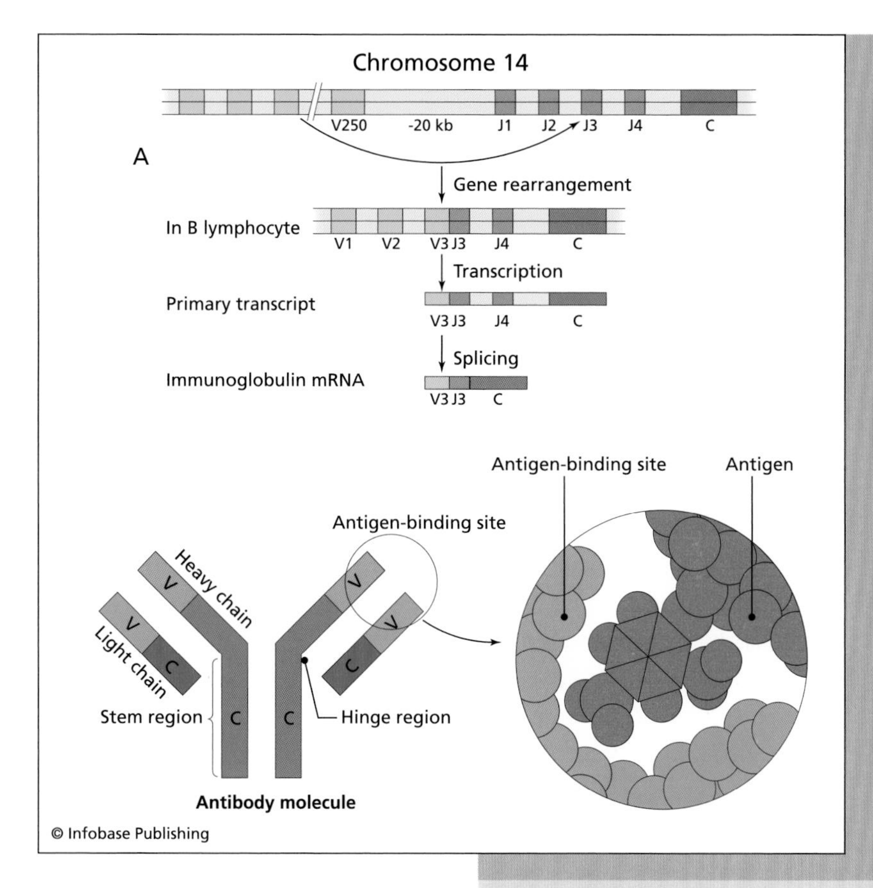

Human antibodies and B cell receptors are made in a complex process that involves a random choice of V, J, and C modules, cutting out and destroying modules that are not used, and rearranging DNA to make a unique gene for each cell (A). When heavy and light chains have been translated into proteins, they are folded and bound into a Y-shaped structure. An opening at the upper tips of the Y provides a binding site for foreign proteins (B).

randomly selected and recombined with one of the four J genes. This pair is then recombined with one of about 200 possible V genes.

Recombination is a process involving several steps in which the DNA strand is cut; regions are discarded or rearranged, and then what is left is reassembled in a new way. Cutting and pasting is the job of a special set of enzymes. In the case of the heavy and light chains, cutting is handled by molecules called RAGs (recombination activating genes), which are only active in white blood cells. V, D, and J genes always lie next to a particular DNA code that is recognized by RAGs. The enzymes bind at these sites and make cuts. A protein machine called the "VDJ recombinase" then comes along to mend the breaks. As it glues the ends of DNA together, the sequences that used to lie between the randomly selected genes are discarded. This puts the remaining variable genes next to the sequence for the constant region. When the RNA-making machinery comes along, it transcribes the pieces as a single gene. This produces a heavy chain immunoglobulin RNA, which has the regions VDJC.

In the meantime, the light chains are being produced from sequences on the second human chromosome, which holds one type of light chain, or the 22nd chromosome, which holds the other. The process resembles the construction of the heavy chain, except that light chains lack D regions, so when they are finished they contain the subunits VJC.

When the RNAs for the two parts have been translated, the cell assembles the new, unique immunoglobulin proteins into a Y-shaped group containing two identical heavy chains and two identical light chains. This leaves a small gap between each heavy and light chain, at the top of each branch of the *Y*. The gap is the docking site for potential antigens. Each of the billions of possible combinations of heavy and light chains has a unique structure, which gives the opening a unique shape. Because an antibody has two identical arms, it can bind at least two copies of the antigen. These may be located on the surfaces of different copies of the foreign microbe, so an antibody acts as a sort of glue, sticking invaders together in large groups that can easily be recognized and digested by immune system cells.

The five types of heavy chains create five major kinds of antibodies that behave in different ways. Some are secreted from the cell; they float freely through the bloodstream until they find a matching antigen. Others remain bound to the surface of the B cell and act as receptors. If they are activated by a T helper cell, the B cell receives a signal to finish developing and reproduce. It begins to produce massive amounts of the antibody. There is no more recombination of the heavy and light chain genes; now every antibody that is made is identical. The cell secretes the copies, which go on to tag foreign antigens. This summons T cells and other types of white blood cells that can kill the invader. The types of antibodies and examples of their functions are listed in the following table.

TYPES OF HUMAN ANTIBODIES		
Type	Location	Function
IgA	Blood, saliva, tears, milk, and bodily secretions	Works in the mucus and stops microbes from attaching themselves
IgD	Surface of B cells	Acts as a receptor on B cells and helps activate them
IgE	On mast cells and other cells involved in tissues that come in contact with the environment and trigger the innate immune response	Helps launch innate immune defenses by releasing histamines
IgG	Floats in the bloodstream	Tags antigens on the surfaces of foreign organisms or infected cells so that they can be recognized and digested by macrophages
IgM	Surfaces of B cells and free floating	Acts as a receptor on B cells and is released in huge numbers when B cells reproduce. This is the first type of antibody produced during infections, and one of its effects is to glue together invading cells in large clumps

T cells have T cell receptors rather than antibodies, but they are made in a similar way and have similar functions. The receptor is also made by recombination of V, D, and J genes. Instead of heavy and light chains, it has two components called beta- and alpha-chains. The beta-chain is made of a randomly chosen V, D, and J gene, while the alpha is similar to the light chain, with only V and J subunits. The receptor remains bound to the T cell, and it recognizes fragments of antigens in combination with MHC proteins.

USING ANTIBODIES IN RESEARCH AND MEDICINE

In 1993 Serge Muyldermans and a team of researchers at Vrije University in Brussels, Belgium, had a lucky accident that may lead one day to promising new therapies involving antibodies. While performing an experiment, they used camel blood instead of serum taken from a mouse. The experiment had a strange outcome that did not make sense until the scientists took a close look at the camel's antibodies. They discovered that the molecules had an unusual structure: They only contained two heavy chains, rather than the heavy and light chains found in other mammals. Further study showed that this is also true for species such as llamas and alpacas, which are closely related to camels through evolution. This is interesting because antibodies have many applications in research and medicine, and in some of these cases camel antibodies may work better than those of other animals.

Their simpler structure makes them smaller, which means they can move through tissues more quickly. They are also more stable when subjected to heat. Several laboratories in Europe, the United States, and elsewhere are, therefore, beginning to use the molecules in their work. For example, Ulrich Wernery, a scientist at the Central Veterinary Research Laboratory in the Middle Eastern country of Dubai, is currently using them to make antidotes to snake venom. Currently, most such "antivenins" are made by exposing horses to small amounts of venom

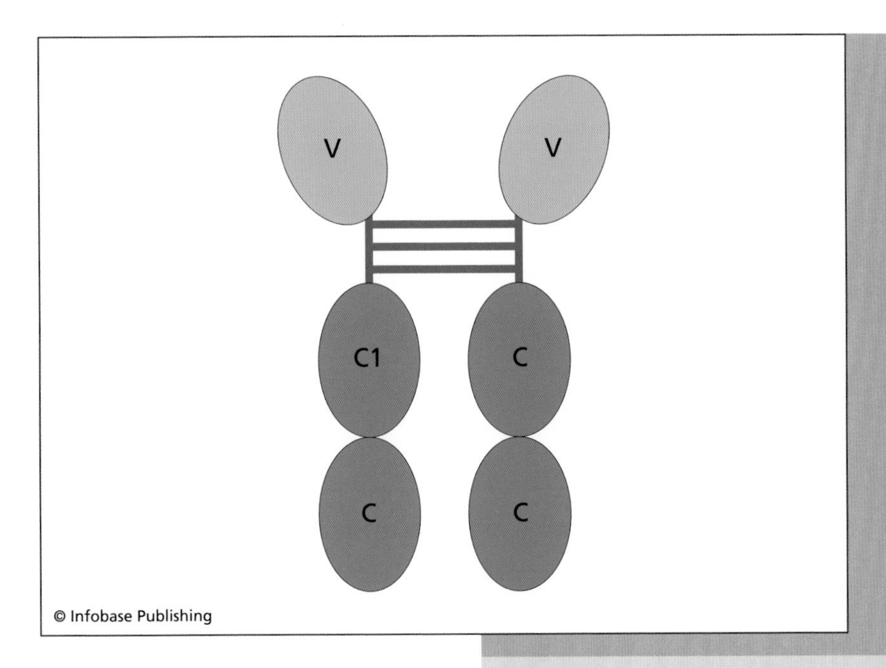

Camels and closely related animals such as llamas produce a unique type of antibody that is missing the light chain. Researchers hope that the molecules can be used in a number of medical applications, such as the production of antivenins, which will avoid some of the problems people have with normal antibodies.

and then extracting the antibodies that the horses make. Some people, however, have allergic reactions to these substances; the therapeutic antibodies themselves are regarded as foreign and rejected by the immune system. Antivenins made in camels are presently scheduled for clinical trials, and Wernery predicts that they will not provoke allergies because of their unique structure.

Camel antibodies are only the most recent in a long line of medical and biological applications of antibodies. One of the main uses is to detect diseases in patients, a process known as serology. Doctors take a sample of blood and measure whether it contains antibodies against particular infectious agents. High amounts of IgM antibodies are a sign that a person is infected with a particular virus or has been infected very recently. Antibodies are also used to look for signs of hepatitis or other liver diseases, including some autoimmune conditions. In pregnancy tests antibodies are used to check a woman's body for hormones produced by an embryo.

Antibodies contribute to a problem that arises during pregnancy when mothers and fetuses have differences in their blood involving Rhesus factors. The blood cells of some people produce the Rh protein, which appears on their surfaces (such people are called "Rh positive"), and the cells of Rh-negative people do not. This can potentially lead to disaster because the mother's body might build antibodies against the Rh proteins of her fetus. While the blood systems of the mother and child are kept separate during pregnancy, the baby's blood sometimes enters a mother's bloodstream during complicated pregnancies or childbirth.

If the fetus has Rh-positive blood, the protein acts as an antigen, and the blood stimulates an immune response. This can cause severe problems in both the mother and child. It also affects the next pregnancy; if the mother's body has antibodies against Rh, they can reach the placenta and attack the blood cells of the fetus. This problem was the cause of many deaths before scientists understood it. Now, doctors routinely screen the blood of mothers and administer anti-RhD treatments before the mother can produce antibodies. Even just a few decades ago, many states required blood tests before issuing marriage licenses in order to warn partners who were at risk of having such babies. Today's easy access to treatment has led nearly all states to drop the requirement.

Another application of antibodies involves giving them to patients when their own bodies fail to produce them in response to a disease, or when the body cannot make them fast enough to respond to an infection or a poison. This procedure is called passive immunity. The strategy is to transplant antibodies that have been made in another person (or an animal such as a horse) to someone suffering from an illness. Examples include antivenins against snakebites and early treatments for rabies. Passive antibody therapies have also been used to treat some types of cancer, multiple sclerosis, and rheumatoid arthritis. When the imported antibodies dock onto foreign substances, they summon macrophages and other white blood cells that wipe out the invader. This part of the immune response works the same as if the antibodies had been made by the patient's body. Yet, the introduction of foreign antibodies sometimes causes problems such as allergic or immune reactions; antibodies made in another body may be regarded as foreign. Additionally, unlike antibodies that develop naturally, the immune system does not develop memory cells, so a patient is not protected from repeated infections.

The fact that antibodies bind to very specific targets has made them valuable in research. They can be used as tools to track, identify, or purify proteins. These techniques require huge numbers of antibodies, which have to be made in cells. Usually the procedure begins by exposing an animal such as a rabbit, horse, or chicken to an antigen, the molecule that scientists would like to tag or study. The animal builds antibodies against the substance. The B cells are extracted, but before their antibodies can be harvested, extra steps are necessary. Just as one key might fit two doors, some antibodies recognize more than one antigen. Additionally, the body often builds more than one antibody against a specific antigen (two different keys that fit the same door). Either of these situations can confuse experiments or cause problems during therapies, so scientists had to find a way to isolate identical and extremely specific antibodies. This required isolating an animal cell that made the best possible antibody, then having that cell copy itself. This creates *monoclonal antibodies,* molecules that are the same because they come from a single parent cell.

The technique for producing this type of antibody was invented in 1975 by German biologist Georges Köhler (1946–95), his professor César Milstein (1927–2002) an Argentine working at the University of Cambridge in England, and the Danish immunologist Niels Kaj Jerne (1911–94). Their discovery led to their sharing the Nobel Prize in physiology or medicine in 1984. One notable problem they had to overcome was the fact that antibody-producing cells do not survive or reproduce well in the test tube. Their solution involved fusing the cells with a type of laboratory-cultured cancer cell called a myeloma. This cell grows readily in the test tube, reproduces very quickly, and is

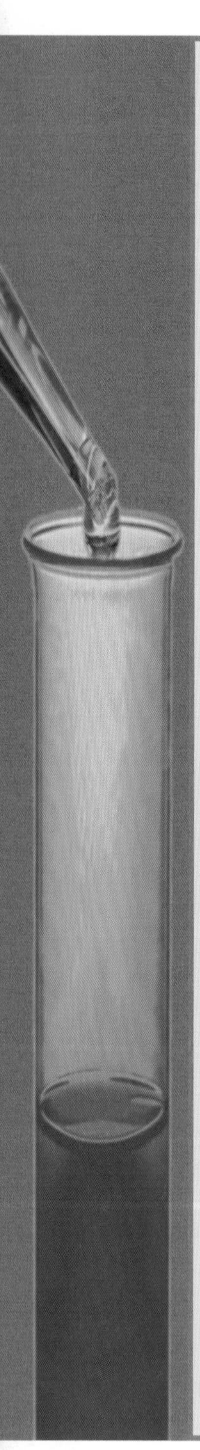

A Devil in Distress: Can Cancer Be a Transmissible Disease?

It is not often that a cartoon character steps up to shed light on the immune system, cancer, and potential new therapies involving T cells. But the Tasmanian devil has suddenly caught the interest of scientists across the world. The real marsupial lives only on the island nation of Tasmania, where its odd shrieks startle tourists and keep people awake at night. Several hundred years ago the devil was also widespread on mainland Australia, but it became extinct there with the arrival of the dingo. Now it faces another threat. In 1996 scientists discovered an epidemic of cancer that was killing the animals in a northern region of the island. Since then the disease has spread so rapidly that researchers estimate it may have decimated half the entire species.

Tumors arise in the animal's mouth and on its face, eventually making it impossible or too painful for the devil to eat, and it starves to death. What has been so puzzling about the disease is that it seems to spread directly from animal to animal. The devils often have fierce fights in which they bite each other on the face, which explains how cancer cells might spread. But while cancer sometimes follows on the heels of an infectious disease (discussed in the next section), it is almost never directly transmissible. Cancer cells are not parasites; they arise through mutations in a person's cells and cannot transmit the disease on their own. If cells from a cancer patient were accidentally transplanted into another person, the immune system of the new host would wipe them out immediately. Only one exception was known, a rare sexually transmitted disease in dogs—and now perhaps in the Tasmanian devils.

In 2006 two Tasmanian scientists, Anne-Maree Pearse and Kate Swift, examined cells from animals with the "devil-facial tumor disease." They discovered some extremely odd characteristics: The cancer cells were missing five of the normal devil chromosomes and had four extra ones. These particular aberrations had never been seen before in the animal's cells. Cells can undergo such massive changes, but the chances of it happening more than once are extremely small. As tumor cells taken from animals all over the country had the same features, it virtually proved that all of the cases could be traced back to one source, rogue cells that evolved in a single animal. Like parasites, they were being passed from one devil to the next in a bite.

The Tasmanian devil in defensive stance, at Tasmanian Devil Conservation Park, Tasman Peninsula *(Wayne McLean)*

Normally the immune system treats tumor cells like any other type of transplanted tissue and rejects them, but even other types of tissue transplanted between devils were not being rejected. Pearse and Swift proposed that this was due to heavy inbreeding within the population. A 2007 study by groups from Australia and Tasmania confirmed this and showed what aspect of the immune system was at fault. Existing devils produce very few types of MHC molecules ("vacuum sweeper" proteins on the surfaces of cells that combine with fragments of foreign molecules and are recognized by T cell receptors). Most species have

(continues)

(continued)

a wide variety of MHCs, which means that two animals are unlikely to have the same ones. This protects them from accidentally taking in diseased cells—such as cancer—from other members of their species. The devils are closely related to each other. It is strong evidence that the animals that live in Tasmania descend from a very small population that arrived on the island a few hundred years ago.

"immortal"; in other words, it goes on copying itself for generation after generation without becoming old or dying. Samples of the antibodies produced by each cell culture are extracted and tested to find one that binds to the antigen the best. It is then selected to produce the finished molecule.

In recent years genetic engineers have developed new techniques to make antibodies by inserting their genes into bacteria, yeast, and other microorganisms to turn them into antibody-production factories. These molecules remain one of the most important tools in biological and medical research.

VIRUSES AND CANCER

The cancer that is wiping out Tasmanian devils is so unusual because it is contagious. In the early 20th century Peyton Rous (1879–1970), a scientist at the Rockefeller Institute for Medical Research in New York, made a discovery that was equally startling for his time: He proved that a type of cancer could be transmitted from one animal to another by a virus.

Rous was working with chickens, which commonly suffered from tumors called "sarcomas," which grow in tissues such as bone and muscle. After watching sarcomas spread like infections through rural farmyards, he began searching for the agent that was transmitting the disease. He removed a sarcoma from the leg of a chicken, prepared and filtered it until there were no more cells, and then injected what remained into a healthy bird. The new hosts developed the same types of tumors. When Rous extracted some of their cells and passed them along to another round of birds, the same thing happened.

Because he had carefully filtered the extracts before injecting them into a new animal, Rous hypothesized that the agent had to be something smaller than a cell—a virus. Most scientists found this very hard to believe; it seemed to contradict the well-accepted theory that cancer arose spontaneously from defects in cells, proposed by the German physician and researcher Rudolf Virchow (1821–1902) in 1863. Even Rous became frustrated and abandoned the line of work after failing to find cancer-causing viruses in mice. Two decades later, however, a colleague asked for his help in the study of warts that arose in mice because of a virus; the warts frequently developed into cancer. The idea of cancer-causing viruses gradually became accepted, and finally, in 1966 Rous was presented with the Nobel Prize in physiology or medicine.

Why does the Rous sarcoma virus cause cancer? The answer lies with a gene carried into the cell by the virus. It encodes a protein called Src (known as v-Src) whose structure is almost identical to a cellular protein (c-Src). C-Src has an important function in a signaling pathway that tells the cell when to divide. Its structure allows it to act as a switch that can be turned on and off by other molecules. In an infected cell these signals are intercepted by v-Src, which has a slightly different structure. It is permanently stuck in the "on" mode, so it permanently broadcasts the signal for the cell to divide, causing a tumor. Several other types of cancer have since been linked to viruses. Many of these do so in a similar way, by interfering with signaling pathways. Nonetheless, the majority of tumors arise from spontaneous mutations rather than infections.

Scientists have not yet developed vaccines that directly target tumors, except in a few highly experimental studies. Theoretically, if a unique protein can be found on a cancer cell—one that is not produced by other cells—it might be possible to manipulate T and B cells to attack the tumor the way they attack parasites. This is difficult because cancer arises from the body's own cells, and the molecules it produces are not truly foreign. But they may have unique characteristics, such as changes caused by mutations, that will allow the body to recognize and target them without also harming healthy molecules by mistake.

On the other hand, there have been several successes in blocking viruses that lead to cancer. The year 2002 saw the development of a vaccine against four forms of the human papilloma virus (HPV), which frequently causes cervical cancer in women. Doctors routinely use the Pap test, or smear, as a diagnostic tool for a woman's risk of developing cancer. In 2006 the virus was the most common sexually transmitted disease in the United States; the Centers for Disease Control estimated that 20 million Americans were infected, many of whom developed cancer. In 2006 the U.S. Food and Drug Administration approved the introduction of the vaccine for general use by doctors.

THE MOLECULES OF AIDS

The World Health Organization estimates that since its discovery in 1981 AIDS has caused more than 25 million deaths around the world, and about 33 million people are currently infected by HIV, the virus that causes AIDS. It is transmitted through blood or body fluids that pass from one individual to another during sex, when they share needles, or from mother to child during childbirth. Researchers have not yet been able to create an effective vaccine against HIV because of the way HIV infects cells, the fact that it targets cells of the immune system that could help fight off infections, and its rapid rate of mutation.

Viruses gain entry to cells by binding to receptor proteins on their surfaces. HIV targets white blood cells such as macrophages and helper T cells; they have a receptor protein called CD24 that a protein on the surface of the virus (gp120) can dock onto. Other proteins jump in to help the surfaces bind and move viral molecules into the cell.

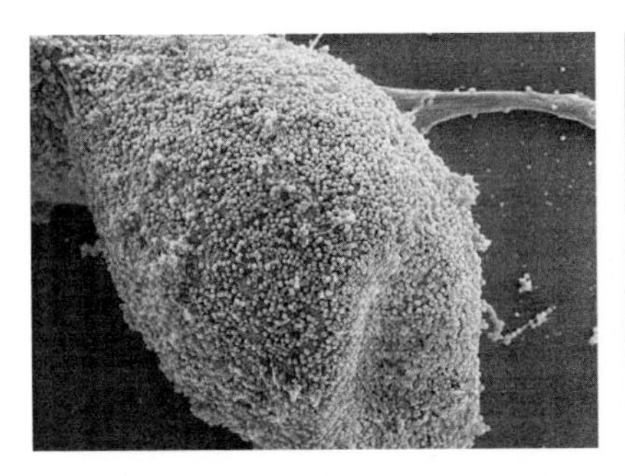

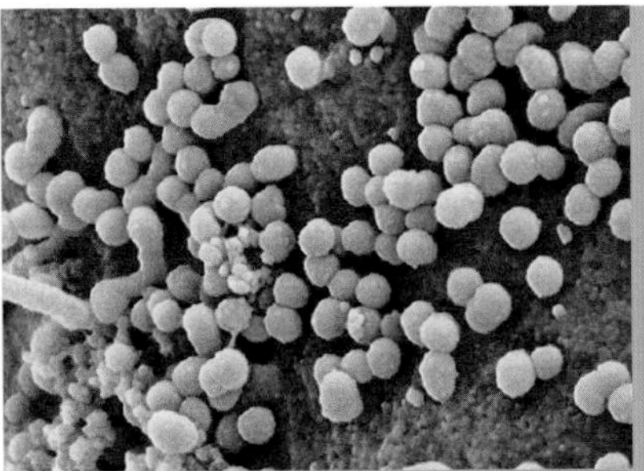

An electron microscope image of HIV docking onto a T cell, the first stage of infection *(NIH)*

HIV contains its genetic material, made of RNA, and a reverse transcriptase—a molecule that can read the RNA code and use it to build a DNA molecule. (This process, described in chapter 2, is the reverse of normal transcription, in which DNA sequences are used to make RNAs.) The new DNA enters the cell nucleus where another protein tool from the virus inserts it into the cell's genome. It stays there permanently and is occasionally used to make the building materials for new copies of the virus. They go on to infect other cells, a process that may take several years. Since the virus infects important types of white blood cells and eventually destroys them, at some point the immune system is so weakened that it can no longer protect the body from other infections. Scientists did not discover AIDS for many years because the victims usually died of strange, secondary diseases rather than from direct effects of the virus.

One reason that it has been impossible to make an AIDS vaccine is that HIV undergoes mutations very rapidly, at a much faster pace than cells or many other viruses. The machinery that copies a cell's DNA has sophisticated safeguards to prevent mistakes, but the HIV reverse transcriptase does not have these quality-control devices, so it constantly makes mistakes as it turns viral RNA into DNA. Most of these changes damage the virus so that it is no longer infectious, but a few get through and make it hard for the immune system to respond to HIV. The body builds antibodies against gp120 and other molecules that appear on the surface of the virus, but mutations cause

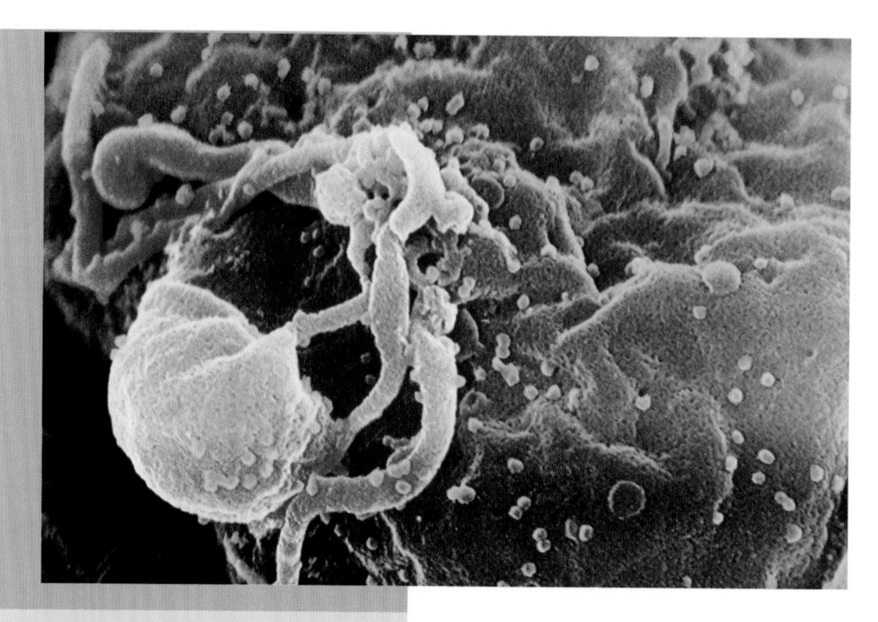

A scanning electron micrograph of a new copy of HIV-1, budding from a cultured lymphocyte in which it has just reproduced. Multiple round bumps on the cell surface represent sites where other copies of the virus are being assembled and budding. *(C. Goldsmith, Center for Disease Control Public Health Image Library)*

small changes in their shapes and chemistry that make them unrecognizable again. For the same reason, it is hard to build an effective vaccine. Several attempts to teach the body to build antibodies against gp120 have been unsuccessful.

Another strategy to fight the virus is to interfere with the stages by which it reproduces. Cells do not have reverse transcriptases, which the virus needs to survive, so current HIV therapies use two drugs that block

(opposite page) HIV can infect humans because proteins on its surface bind to some types of white blood cells and gain entry. Once inside, the virus writes its RNA into DNA and inserts it into the cell's genome. At some point the cell begins using the new genes to produce viral RNA and proteins, which are used to make new copies. The components have to be properly made and assembled or else the copies cannot infect new cells. The most effective current therapies interfere with this process by blocking the activity of HIV molecules.

their activity. Still other strategies are based on the fact that, at one stage in building the virus, three of its proteins must be cut at precise places. The fragments have to move to different parts of the virus to give it the proper architecture. A third anti-AIDS drug blocks the activity of the molecule that makes the cuts. In combination, these drugs have proven to be very effective in fighting most strains of HIV. With long-term treatment, many patients no longer have symptoms or show traces of the virus.

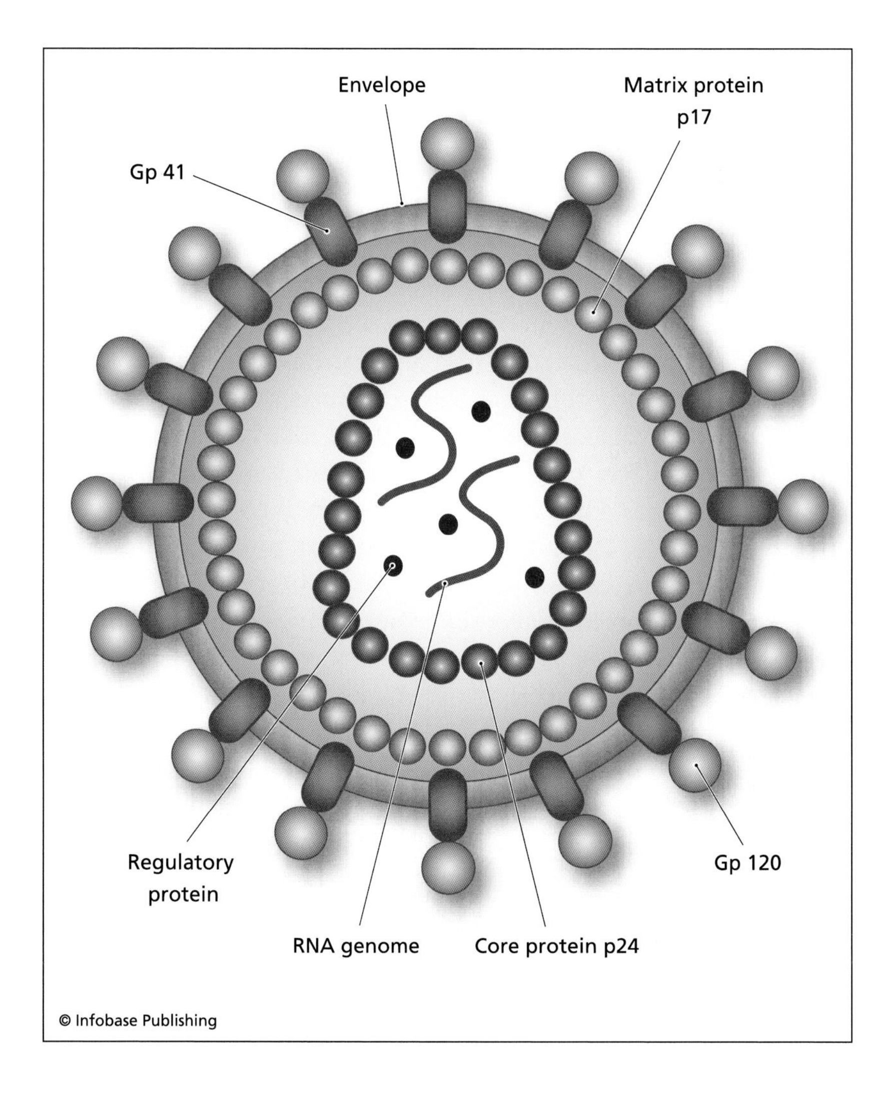

AIDS researcher Robert Siliciano of Johns Hopkins University School of Medicine, says this treatment is not a definitive cure, however. It probably does not completely eliminate HIV, which can lurk quietly in the cells for many decades. People with the disease may have to be treated their entire lives unless new treatments are found, and most people affected by AIDS, including those living in the hardest-hit regions of the world, do not have access to the "cocktail" of drugs because they cannot afford them. Furthermore, HIV evolves so quickly that some strains have already become resistant to the drugs. For all of these reasons, AIDS is likely to remain a long-term threat to humankind until there is a breakthrough in the creation of a vaccine or another treatment.

AMYLOIDS AND NEURODEGENERATIVE DISEASES

Alzheimer's disease is now recognized as the most common form of dementia—memory loss and a progressive loss of mental abilities—among senior citizens. Scientists still do not know exactly how the disease develops or why it strikes some people in their 60s and 70s while others lead healthy mental lives until an advanced age. The main problem seems to be a structural change in a molecule that misfolds, collects in dense clumps of fibers, and interrupts communication between brain cells. Similar clusters of proteins have been found in several other neurodegenerative diseases.

Chapter 1 introduced Alois Alzheimer, the German researcher who discovered the disease. He announced the condition in 1906 in a paper called "A Peculiar Disease of the Cerebral Cortex," given at a medical conference in the southern German town of Tübingen. Within a few years other physicians had found other cases and begun referring to it as Alzheimer's disease. Alzheimer was gaining a reputation as an expert in the study of how the structure of the brain changed during syphilis and other diseases. The case he described involved a patient named Auguste Deter, whom he had met several years earlier while working at the State Institute for the Mentally Ill and Epileptics in Frankfurt. In 1901, when Deter was 51 years old, she was admitted to the institute after her behavior at home became increasingly difficult for the family to deal with. She had memory problems, was having trouble speaking and understanding what was said to her, and had become paranoid, claiming that her husband was having affairs. She was admitted to the institute, where her condition became progressively worse. Alzheimer examined her several times and recorded some of the symptoms: "While she was chewing meat and asked what she was doing, she answered that she was eating potatoes and horseradish. When objects are shown to her, she forgets which objects have been shown after a short time. She constantly speaks about twins. . . . Asked to write Auguste D., she tries to write Mrs and forgets the rest. It is necessary to repeat every word."

In 1906 the German physiologist Alois Alzheimer gave a talk in which he described a neurodegenerative disease marked by accumulations of amyloid plaques in the brain. Alzheimer's disease is now recognized as a leading cause of dementia and death among the elderly. *(Ron Moody)*

After her death in 1905 Alzheimer obtained her brain and examined it closely. He discovered that the space between cells had filled up with "plaques," tangled mixtures of starches and protein fibers called "amyloids." Presently, these accumulations are thought to be toxic to neurons, eventually causing their death. In 1985 Konrad Beyreuther of the University of Cologne, Germany, and a group of Australian researchers purified the major protein

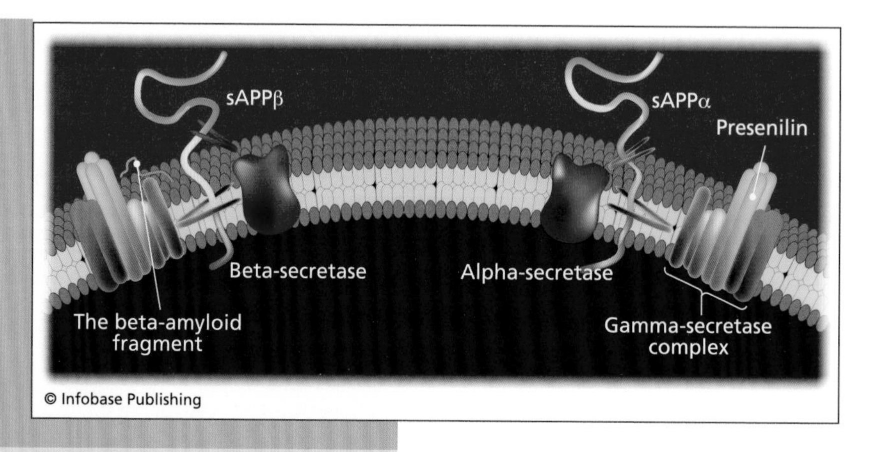

Alzheimer's disease is usually accompanied by a buildup of the protein beta-amyloid, which accumulates as insoluble clumps in the space between brain cells. Beta-amyloid is a fragment of a larger protein, APP, found in the membranes of neurons. It can be cut in at least two ways: (*left*). If APP is cut by membrane proteins called the gamma-secretase complex and beta-secretase, the result is beta-amyloid; (*right*) if, instead, it is cut by alpha-secretase, the fragment that is produced is harmless and, in fact, probably helpful for the normal function of the brain.

from the plaques. Over the next few years the molecule, called beta-amyloid, was identified as a fragment of a much larger protein normally found in the membranes of nerve cells: APP (for "amyloid precursor protein").

A great deal of work has gone into investigating APP in hopes of discovering how and why cells make the fragment. Beta-amyloid is only produced when molecules called "beta-secretase" and "gamma-secretase" cut APP in specific places. First beta-secretase slices off a part of the molecule outside the cell, much like trimming a bush. Then gamma-secretase gets access to a region of the protein that lies inside the membrane and cuts it there. This uproots the fragment, and it begins to accumulate between cells.

The transmembrane region of APP can also be cut by a third molecule, called "alpha-secretase." This cut produces a fragment that does not accumulate and is not dangerous. Instead, it seems to be helpful in the formation of contacts between nerve cells and in the processes of memory and learning. APP's jobs in the healthy brain are not well understood, but the fragment produced by alpha-secretase seems to be important.

A great deal of ongoing Alzheimer's research focuses on discovering why beta- and gamma-secretases get the upper hand over alpha-secretase. All of the molecules, including the beta-amyloid fragment, surely have important healthy functions, but most of them are still unknown. The fragment itself is probably completely harmless until it accumulates in a certain way, and to do so it may need the help of yet other molecules. Part of the problem may be that it folds the wrong way. Researchers have noted that many diseases—including prion diseases (discussed in the next section), cystic fibrosis, and sickle-cell anemia—arise when proteins are improperly folded, which changes their functions.

Alzheimer-like plaques are found in many other diseases that affect the brain, including Huntington's disease (where they clog up the cell nucleus) and type 2 diabetes. At least a dozen different proteins—usually one for each disease—can form plaques. The fragments that make these clumpy fibers often have a fairly chaotic structure until they begin to accumulate, then they form long, orderly rows of beta sheets that cannot be dissolved. Preventing Alzheimer's and similar diseases may depend on understanding each stage of the production, misfolding, and accumulation of the fragments, and then to find substances that block some of the steps.

PRIONS AND MAD COW DISEASE

In the late 19th century Louis Pasteur (1822–95) became firmly convinced that all infectious diseases were caused by bacteria. The idea became so widely accepted that when a Frenchman named Charles-Louis-Alphonse Laveran (1845–1922) claimed that malaria was caused by another type of one-celled parasite, it took several years for the medical community to believe him. (When they did, he was awarded the 1907 Nobel Prize in physiology or medicine.) Early virus researchers confronted a similar situation. And very recently another scientist had a similar experience when he claimed that mad cow disease and some related infections were caused by the transmission of a prion, an unusual form of a protein, rather than viruses or bacteria.

Mad cow disease and other transmissible spongiform encephalophathies (TSEs), resemble Alzheimer's disease, because patients also develop accumulations of protein fibers in the brain. Holes form in the tissue, and the brain takes on the appearance of a sponge. Once the first symptoms appear, the disease develops more quickly than Alzheimer's and many other neurodegenerative diseases. The TSE that infects sheep is known as scrapie, because infected animals begin to behave oddly and scrape the wool off their skin. In the 1980s cows in Great Britain developed the disease; the cause was traced to meal they had eaten that contained ground-up remains of sheep. Since then several humans have become infected and died from the disease, very likely because they ate the meat of infected cows. It has taken a huge effort on the part of the government and medical officials to try to control the disease by interrupting the food chain and removing infected meal from animals' diets.

A similar human disease had been discovered in a tribe in New Guinea in the 1920s. The native people called it *kuru,* or "shaking disease," because of the behavior of infected people. Medical researchers eventually traced the cause to cannibalism; the tribe had a ritual of eating their dead, particularly the brain. All the victims had participated in the rituals and eaten infected human tissue. In the 1950s the tribe stopped the practice, and the disease gradually began to die out; however, scientists believe it can "incubate" for decades in a person's body before the symptoms arise. Cases in New Guinea continued to be found into the 1990s.

Until the 1960s the infectious agent responsible for TSEs remained a complete mystery; no bacteria, virus, or other parasite had been found. Then Tikvah Alper (1909–95), head of the Medical Research Council Radiopathology Unit at Hammersmith Hospital in London, made the revolutionary proposal that a protein might be responsible. She took infected material from an animal and exposed it to high doses of ultraviolet light. Even though this procedure destroyed DNA or RNA in the sample, animals could still catch the disease through food. Viruses, bacteria, and other known infectious agents relied on these molecules to infect the cell. The only thing left, Alper concluded, was a protein.

This was at least as radical a claim as Rous's hypothesis that viruses could cause cancer, and most researchers did not believe or accept her conclusions. It took nearly 20 more years for the laboratory of Stanley Prusiner (1942–), a neurologist and biochemist at the University of California, San Francisco, to purify the disease-causing substance. He claimed that it was a protein, and coined the term *prion* to describe it. The discovery was greeted with great skepticism, despite the fact that Prusiner was his own hardest critic and had tried every method he could think of to disprove what his experiments seemed to be saying. Prusiner and his colleagues needed two more years to identify the gene encoding the molecule, which he called PrP. The abbreviation stands for "protease-resistant protein," in other words, a molecule that cannot be broken down by the cellular enzymes that normally take apart molecules. For his work on prions, Prusiner was awarded the 1997 Nobel Prize in physiology or medicine.

Prions seem to cause disease when something changes their normal secondary structure, which consists of alpha helices, into a form containing beta sheets that the body cannot break down. Mutations that change the shape of PrP sometimes occur naturally—at a very rare rate—and are thought to cause human Creutzfeld-Jakob's disease (CJD). But most cases seem to occur when a person eats meat that contains misfolded prion proteins. Somehow this triggers the person's own PrP to change its shape and begin to form fibers between brain cells.

Like APP, the culprit in Alzheimer's disease, PrP is a membrane protein found on the surface of brain cells as well as other types of cells throughout the body. Its normal biological activities are not yet known, but it seems to play a role in establishing long-term memory (another similarity to APP). The fact that PrP strongly binds to copper ions may be another clue to its functions. In the body, copper plays an important role in enzymes, in obtaining energy for cells, and in the function of nerves.

PrP probably becomes misfolded when it binds to another molecule. In 2003 several laboratories discovered that the protein docks onto nucleic acids, the building blocks of RNAs and DNA, hinting that this interaction might be important in how the protein folds and functions.

STEROIDS AND OTHER HORMONES

Most people have heard of steroids through the news; barely a day goes by without another revelation that a professional athlete has used them or other performance-enhancing substances. Steroids are artificially created molecules that have powerful effects on the body because they closely resemble testosterone, one of the body's own hormones.

Dozens, possibly hundreds, of types of hormones are produced and secreted by the body's glands and cells. They spread quickly through the bloodstream and stimulate rapid, powerful responses from many types of cells. Testosterone, for example, acts as a strong developmental signal during male puberty that guides the development of the sex organs, deepens the voice, and promotes the growth of bones and muscles. This explains why synthetic hormones can increase the size of athletes' muscles, but their use often has other undesirable effects, including giving men female characteristics such as enlarged breasts. Females also produce testosterone, but their bodies convert it into the estrogen hormone estradiol. Women who take steroids may develop deeper voices, facial hair, and other features of males.

The discovery in the 1930s that hormones control crucial aspects of female reproduction led to the development of the birth control pill, whose main ingredients are the hormones estradiol and progesterone. They function by tricking a woman's body into thinking it is pregnant, blocking the monthly release of an egg and changing her body in other ways to prevent pregnancy.

Hormones work by docking onto receptor proteins and changing the activity of genes, either by triggering signaling pathways or more direct means. Some are able to slip directly into the cell through the membrane. This makes them ideal messengers to control major body systems, such as blood pressure, temperature, growth, immune responses, emotions, and digestion.

Hormones are made in cells or glands scattered throughout the body, including regions of the brain, the thyroid gland, the pancreas, and sex organs. The tissues that secrete them are called endocrine glands, which simply means that the substances they produce stay in the body rather than leave it. Small groups of cells within other organs can also secrete hormones. They enter the bloodstream and travel through the body until they recognize receptors on their target cells. Steroid hormones, such as testosterone and estradiol, are made from cholesterol. Other hormones, such as insulin, are small proteins. The amines are made from tyrosine, one of the amino acids. Proteins and some of the amines have to be packed into small membrane compartments to be released from cells and absorbed by others.

Because of their widespread and rapid action, hormones evolved important functions such as control of the "fight-or-flight" response, the body's first reaction to frightening or threatening situations. This reaction is governed by the hormone adrenaline (also called epinephrine), which increases the heart rate and reroutes glucose (which provides energy) from unimportant cells and systems, making as much as possible available to muscles.

Hormones and the endocrine system play such important roles in the body that defects or problems often lead to serious diseases, such as the following:

- Hypothyroidism means that the thyroid gland is not active enough and thus does not produce enough thyroid hormones such as thyroxine. These molecules play an important role in regulating the body's metabolism. The condition is often caused by inflammations of the thyroid gland. It may also be the result of surgery that has removed part of the gland, as, for example, in thyroid cancer cases. It can be treated effectively in most patients by giving them extra hormone.

- Hyperthyroidism occurs when the thyroid gland is too active and produces too much thyroxine or other hormones. The result is often excessive growth and hyperactivity among the body's organs. The disease often follows autoimmune diseases in which the body raises antibodies against proteins on cells in the thyroid. These activate the tissue to produce more hormones. Treatments include drugs that reduce the output of the thyroid gland, radioactive iodine (which only kills thyroid cells), or surgery.
- Diabetes mellitus type 1 is an autoimmune disease that destroys cells in the pancreas that produce the insulin hormone. It must be treated with injections of insulin, or else it is almost always fatal. Formerly thought of as a childhood disease, an increasing number of cases are now found in adults. The causes are not known, but researchers believe that there may be several routes to developing the disease, including viral infections, certain chemicals, and even the body's building antibodies against proteins from cow's milk.
- Diabetes mellitus type 2 is a type of insulin "resistance" in which the body often produces normal amounts of insulin, but cells do not respond to it properly. Glucose levels rise in the blood, because the liver cannot store it and muscles do not take up enough of it. Symptoms are fatigue, weight gain, and high blood pressure, and the result can be permanent damage to organs. The disease can be partially treated by changes in diet and increasing the amount of exercise. In the past the disease usually affected adults, but an increasing number of young people have been diagnosed with it, probably due to changes in diet and a reduced amount of exercise.
- Addison's disease is the result of defects in the adrenal gland, which produces steroid hormones. The condition may arise for genetic reasons, because of problems in the way the body processes cholesterol, or from damage to

the gland. Symptoms include fatigue, muscle weakness, vomiting, changes in personality, and joint and muscle pains. The disease is usually treated by giving patients extra cortisol.

- Metabolic syndromes are a collection of disorders that are accompanied by a high risk of cardiovascular disease and diabetes. The causes are unclear, but a patient often has several types of hormonal imbalances, particularly insulin resistance and abnormal levels of cholesterol. Symptoms often include obesity and high blood pressure. Treatment usually involves changes in lifestyle—a better diet and more exercise—and may include hormonal therapies.

Treating a hormone insufficiency may be as simple as giving patients extra amounts of the molecule. This has generally become much easier since the invention of genetic-engineering techniques that turn bacteria or other types of cells into factories for human molecules. Other hormonal conditions are harder to treat, but in some cases they can be avoided through a proper diet and exercise.

NICOTINE AND ADDICTION

Nicotine is a bitter-tasting substance found in the leaves of the tobacco plant. When smoked in a cigarette, it passes through the lungs, enters the bloodstream, and reaches the brain in seven or eight seconds. Because its structure is similar to a neurotransmitter called acetylcholine, it imitates this natural signal. When it binds to its receptors, it opens calcium channels and causes the body to release the hormone adrenaline. This raises the rate of the heartbeat and breathing and releases glucose into the blood.

Nicotine also binds to cells that produce dopamine, another neurotransmitter, making them release it in large amounts. This stimulates new regions of the brain and causes feelings of pleasure. Smoking causes excess dopamine to collect in the synapses between nerve cells. Under natural conditions these extra amounts are quickly reabsorbed, but nicotine prevents this. It blocks a protein called monoamine oxidase, the molecule that normally breaks down dopamine and similar neurotransmitters.

The entry of nicotine into the brain triggers the release of several other neurotransmitters, including norepinephrine, vasopressin, arginine, and beta-endorphin. These substances are sensitive to dosages, so the effects of nicotine change as a person smokes more. Low doses usually act as a stimulant; much higher amounts act as a mild sedative, reducing pain and anxiety. The effects last for up to two hours.

When drugs are made, they are rigorously tested to ensure that they have very specific effects. Ideally they act only on the specific molecules, cells, and body systems that need to be treated. Nicotine and the steroids discussed in the last section would make poor drugs, except in very unusual cases, because they affect so many different types of tissues with unpredictable effects.

Nicotine-related problems are only one of the dangers of smoking. Researchers have discovered that cigarettes and other tobacco products contain at least 4,000 chemicals. The most dangerous are tar, which causes cancer and various types of lung disease, and carbon monoxide, which leads to heart disease.

One result of regular smoking and other uses of tobacco is that the brain's "reward pathways" are stimulated over and over again. To experience the same effects, a person has to smoke again. This leads to powerful cravings and addiction and is similar to how people become addicted to illegal drugs such as cocaine or heroin. The body builds a tolerance to them and the brain lowers its own production of neurotransmitters and other substances that stimulate the same pathways. When a person stops taking the drug, they miss the pleasurable feelings that it causes, and the brain no longer stimulates the pathway in its normal way. The absence of these types of stimulation means that a person can begin to experience painful withdrawal symptoms, depression, and irritation within just a few hours after the drug's effects begin to wear off. The feelings may go on for several days or longer. There may be even worse side effects. Most of the neurotransmitters that are affected by drugs have multiple functions in the brain, such as supporting processes like memory and learning. Over the long term, drug use can have a permanent effect on a person's mental skills.

Addiction has both physical and psychological aspects. Even when a person's physical dependency on a substance has worn off, he or she may still be tempted to keep using it because of the behaviors associated with it. Buying cigarettes and smoking them at certain times may have become part of the daily routine and something to do with friends, for example. It is often just as hard to change these habits as to overcome the pain and discomfort of withdrawal.

CONCLUSION

The sophisticated immune systems of humans and other animals have evolved as a dialogue between species and the diseases that arise in their environments. Microbes also evolve as a response to the environment of their hosts' bodies. Studies of genomes reveal that disease-related genes change at a faster pace than most other DNA sequences, because any mutation that helps an animal resist disease can have a major impact on its ability to survive and reproduce.

Modern medicine began with the discovery that microorganisms cause disease and with vaccines and modern drugs that can combat many of them, taking advantage of the way the immune system copes with infectious diseases. But in many of these cases, cures do not yet exist, partly because viruses such as HIV mutate quickly and outwit the body's defenses.

Defeating these sophisticated invaders will require a new type of medicine based on interfering with molecules and processes within cells. The same approach will be necessary to treat today's greatest killers: heart disease, cancer, and neurodegenerative disorders. Many of these problems can be traced to defects in genes or processes within cells that do not cause problems until old age. They are built into the system, so solving them will require adjusting the system by replacing defective molecules and molecular machines.

These measures will not cure all of humanity's health problems, however. Eventually everyone dies. One day medicine will likely rid humanity of its most dangerous diseases; then people will die for other reasons. Over the next century science may be able to raise the average life expectancy considerably. The quality of that life, however, will still depend on healthy behavior. Unfortunately, the same culture that has brought progress in medicine has led to a lifestyle in which people do not eat well or get enough exercise and in which they abuse their bodies with drugs. Obesity and type 2 diabetes, in which exercise and diet play a huge role, affect more people at an earlier age than ever before. The lesson to be learned is that even the most modern medicine will not protect people from themselves.

Chronology

1774	Joseph Priestley discovers oxygen.
1776	Antoine Lavoisier shows that air is a compound gas made of nitrogen, oxygen, and other elements.
	Edward Jenner develops the first vaccine, using cowpox to train the immune system to fight smallpox.
1828	Friedrich Wöhler synthesizes the substance urea, showing that it is possible to make organic substances artificially.
1840	F. L. Hünefeld accidentally creates the first protein crystals of hemoglobin.
1847	Louis Pasteur discovers that crystals of tartaric acid polarize light in different ways, because the substance comes in two asymmetric forms. This suggests that light can be used to study the atomic composition of substances.
1858	Charles Darwin and Alfred Russel Wallace announce the theory of evolution to the world.
	Archibald Scott Couper and Auguste Kekulé simultaneously publish the first diagrams of molecules in which the positions of atoms and their relationships to each other are shown.

	Rudolf Virchow states the principle of *"Omnis cellula e cellula"* (every cell derives from another cell), including cancer cells.
1865	Gregor Mendel discovers fundamental principles of heredity in experiments with peas and other plants.
1894	Emil Fischer describes the fact that specific enzymes recognize and change each other by using the metaphor of locks and keys.
1895	Wilhelm Röntgen discovers X-rays.
1896	Eduard Buchner successfully performs alcoholic fermentation in extracts from yeast, showing that biochemical processes can be carried out without cells.
1900	Hugo de Vries, Erich Tshermak von Seysenegg, and Carl Correns independently rediscover Mendel's laws.
1906–23	Emil Fischer shows that the amino acids in proteins bind to each other by eliminating water molecules.
1908	Archibald Garrod proposes that genes might be related to the activity of enzymes.
1911	Thomas Hunt Morgan proposes that genes are arranged on chromosomes in a linear way, and his group studies linkage—patterns in which genes are inherited together or separately—to plot the positions of specific genes.

1912	Max Von Laue, W. Friedrich, and P. Knipping observe diffraction patterns when X-rays pass through crystals of copper sulfate.
	William Lawrence Bragg discovers that X-ray diffraction patterns correspond to a regular arrangement of atoms in a crystal.
1927	Hermann Muller shows that radiation causes mutations in genes that can be passed down through heredity.
1928	Frederick Griffith changes one type of bacterium into another using "information" obtained from dead cells, opening the door to the discovery of the molecule that genes are made of.
1930	William Astbury uses X-rays to study protein fibers for the first time.
1931	Archibald Garrold proposes that diseases can be caused by a person's unique chemistry; in other words, genetic diseases may be linked to defects in enzymes.
	Linus Pauling publishes "The Nature of the Chemical Bond," which unites quantum mechanics and chemistry.
1934	John Bernal and Dorothy Crowfoot Hodgkin obtain the first X-ray diffraction pattern from a protein crystal.
1935	Nikolai Timofeeff-Ressovsky, K. Zimmer, and Max Delbrück publish a groundbreaking work

on the structure of genes that proposes that mutations alter the chemistry and structure of molecules.

1940 George Beadle and Edward Tatum prove that a mutation in a mold destroys an enzyme and that this characteristic is inherited in a Mendelian way, leading to their hypothesis that one gene is related to one enzyme (protein), formally proposed in 1946.

1944 Physicist Erwin Schrödinger publishes his book *What Is Life?* which reiterates some of the ideas of Timofeeff-Ressovsky, Zimmer, and Delbrück and attempts to analyze the physical nature of genes and other aspects of the cell. The book inspires an entire generation of biochemists and biophysicists.

Oswald Avery proves that genes are made of DNA.

1947 The Medical Research Council of England creates a structural biology unit at the Cavendish Laboratory in Cambridge and a biophysics laboratory at King's College, which will solve the structures of proteins and DNA within about a decade.

1950 Erwin Chargaff discovers that the proportions of A and T bases in an organism's DNA are identical, as are the proportion of Gs to Cs.

1951 Rosalind Franklin and members of Maurice Wilkin's lab at King's College begin experiments with DNA, defining some of the dimensions that the molecule has to fit.

Pauling and R. B. Corey discover the alpha helix secondary structure in proteins.

Barbara McClintock presents her findings on the discovery of "transposable elements" (transposons) in maize.

1952 Franklin obtains exact measurements of the dimensions of DNA's structure.

1953 Pauling proposes a structure for DNA that is quickly shown to be incorrect.

1953 On February 27 and 28, James Watson and Francis Crick finish their model of DNA, which fits all of the requirements from Franklin's data and explains Chargaff's rules. Their paper is published in April, alongside papers from Wilkins and Franklin that support their findings.

Hans Adolf Krebs discovers the steps by which cells metabolize carbohydrates, now known as the Krebs cycle.

1954 Max Perutz discovers that soaking protein crystals in heavy atoms will provide information needed to turn diffraction patterns into structural pictures of proteins. He publishes an initial structure of hemoglobin.

1957 Crick presents the "central dogma," a challenge to the community to discover how the information in genes is transformed into proteins.

1960 John Kendrew obtains the first high-resolution three-dimensional structure of a protein, myoglobin.

Perutz quickly follows with the structure of hemoglobin.

Arthur Kornberg makes a new strand of DNA in a test tube from an existing strand, by adding energy molecules and free nucleotides.

1961 Sydney Brenner, François Jacob, and Matthew Meselson discover messenger RNA as the template molecule that carries information from genes into protein form.

Crick and Brenner suggest that proteins are made by reading three-letter codons in RNA sequences, which represent three-letter codes in DNA.

M. W. Nirenberg and J. H. Matthaei use artificial RNAs to create proteins with specific spellings, helping them learn the complete codon spellings of amino acids.

1962 Watson, Crick, Perutz, and Kendrew receive the Nobel Prize.

1970 Hamilton Smith and Kent Wilcox isolate the first restriction enzyme—a molecule that cuts DNA at a specific sequence—which will become an essential tool in genetic engineering.

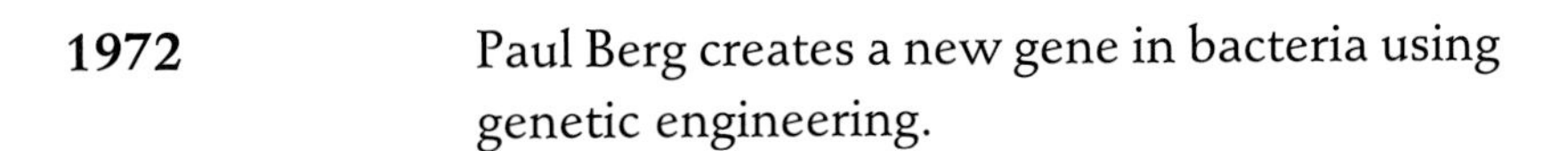

1972 Paul Berg creates a new gene in bacteria using genetic engineering.

1973 Stanley Cohen, Annie Chang, Robert Helling, and Herbert Boyer create the first transgenic organism by putting an artificial chromosome into bacteria.

1975 Edward Southern creates Southern blotting, a method to detect a specific DNA sequence in a person's DNA. The method will become crucial to genetic testing and biology in general.

César Milsein, Georges Kohler, and Niels Kai Jerne develop a method to make monoclonal antibodies.

1977 Walter Gilbert and Allan Maxam develop a method to determine the sequence of a DNA molecule.

Fredrick Sanger and colleagues independently develop another very rapid method for doing so.

Phil Sharp, Louise Chow, Rich Roberts, and Pierre Chambon discover that genes can be spliced, leaving out information in the DNA sequence as a messenger RNA molecule is made.

1978 Kurt Wüthrich and R. R. Ernst use nuclear magnetic resonance to solve protein structures.

1982	Sanger and his colleagues obtain the complete DNA sequence of a virus that infects bacteria.
1985	Kary Mullis develops the polymerase chain reaction, a method that rapidly and easily copies DNA molecules.
1987	The U.S. Department of Energy proposes a project to sequence the entire human genome.
1989	The Human Genome Organization is founded.
1992	J. Craig Venter founds the Institute for Genome Research, and, later, Celera Genomics, which will play a key role in the Human Genome Project.
1995	The Institute for Genome Research announces the completion of the entire genome sequence of a cell: the bacterium *Haemophilus influenzae.*
	The U.S. Food and Drug Administration approves the first protease inhibitor for use in treating AIDS.
1996	A large international group completes the sequencing of the yeast genome.
1997	A sheep named Dolly is the first mammal to be cloned from an existing adult animal.
1998	Completion of the genome of the first multicellular organism, the worm *Caenorhabditis elegans,* is announced.

2000 Researchers at the Human Genome Organization and Celera Genomics announce the completion of a "working draft" of the entire human DNA sequence.

2001 Celera Genomics announces the first complete assembly of the human genome.

Glossary

actin a protein that forms fibers to help create the cytoskeleton, which gives cells their shape. The fibers are used to transport molecules through the cell and to carry out mechanical jobs, such as helping muscles expand and contract.

adenosine triphosphate (ATP) a compound containing three phosphate groups that is broken down by enzymes to provide energy for chemical reactions in the cell

alpha helix a small shape that forms in proteins due to the chemical attraction between neighboring amino acids. Alpha helices and beta strands are secondary structures, the smallest functional shapes in proteins.

alternative splicing a process by which some protein-encoding regions of RNAs are removed to create different forms of proteins based on a single gene

amino acid the fundamental chemical subunit of a protein. There are 20 types, all built around an identical core of carbon, hydrogen, oxygen, and nitrogen atoms and made unique by a side chain of other atoms.

amyloid fiber a collection of dense proteins that collect between brain cells and cannot be dissolved, leading to the death of neurons in Alzheimer's disease and other neurodegenerative conditions

antibody a molecule on B cells that plays a key role in adaptive immunity. Antibodies are produced through random rearrangements of genes, creating a huge range of structures that can recognize foreign molecules or substances. They mark invading viruses or microbes for destruction by immune system cells.

antigen a molecule recognized by an antibody

apoptosis a cellular self-destruct program that helps sculpt tissues as an embryo develops and that destroys some types of diseased cells

axon a long extension growing from a neuron that stimulates neighboring cells

beta strand a small shape that forms in proteins due to the chemical attraction between neighboring amino acids. Beta strands may link to each other to create beta sheets. Beta strands/sheets and alpha helices are secondary structures, the smallest functional shapes in proteins.

chemokine a type of chemical signal that attracts cells

chemotaxis a process by which cells migrate in response to a chemical signal. Often, the cell senses changes in the concentration of a molecule to find its way.

codon a three-letter sequence in RNAs that is used to translate the four-letter alphabet of DNA and RNA into the 20-letter amino acid code

constant region a subregion of identical antibody molecules. Constant regions combine with variable regions to create billions of types of antibodies with unique structures.

cytoskeleton a system of protein tubes and fibers that gives the cell shape and structure and plays a key role in such processes as cell division and migration

dendrite a branchlike network of extensions that grow from neurons and are stimulated by chemical signals from neighboring cells

diffraction pattern an image that is produced when X-rays or other high-energy beams are deflected by the electrons in a molecule. The pattern can be reinterpreted into a three-dimensional map of the positions of atoms within the molecule.

DNA (deoxyribonucleic acid) a molecule made of nucleic acids that forms a double helix in cells, holds a species' genetic information, and encodes RNAs and proteins

DNA transposon a DNA sequence that is cut out of the genome and reinserted in another place by enzymes

domain a functional three-dimensional module in a protein formed when smaller secondary structures fold into a larger unit

endosome a small compartment with a membrane, used to move substances into the cell

exon a region within a gene that encodes a protein or part of one

exon junction complex (EJC) a group of proteins attached to RNAs at sites where they have been spliced. EJC proteins sometimes contain instructions telling the cell where to move an mRNA, and they also help the cell recognize mutations or other defects in molecules.

exosome a small compartment with a membrane, used to move substances out of a cell

flagellum a hairlike tail made of proteins, found on sperm and many species of bacteria

frameshift a change in the "spelling" of a DNA or RNA molecule that shifts the borders of codons recognized by a ribosome as it translates RNA into protein

G cap a structure added to the head of an mRNA molecule to protect it and help in its translation when it leaves the nucleus

guanosine triphosphate a form of the nucleic acid guanine, linked to sugar and three phosphate groups, that provides energy for some cellular reactions

hemoglobin a protein in red blood cells that binds oxygen atoms and carries them to tissues throughout the body

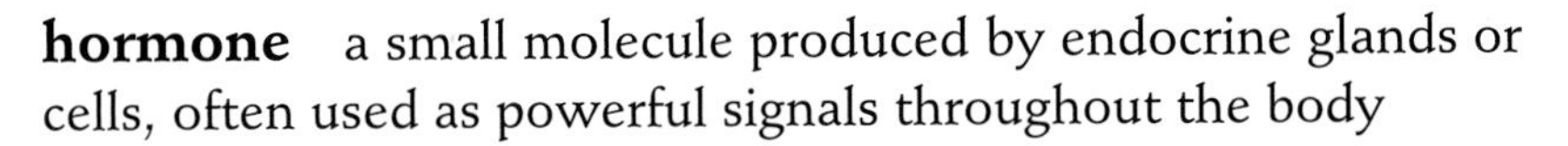

hormone a small molecule produced by endocrine glands or cells, often used as powerful signals throughout the body

intron a sequence in an RNA or gene that does not encode a part of a protein

ion an atom that bears a charge because it lacks an electron or has an extra one

ion pump a protein that pushes ions "upstream" through membranes against the direction they would normally flow, for example, by moving a positively charged particle into a cell that already bears a positive charge

kinase receptor a molecule that activates cell signaling using ATP

ligand a small molecule that binds to a receptor or another protein

lysosome a membrane-enclosed compartment in the cell that contains powerful enzymes used to break down bacteria and to recycle worn-out cellular structures

major histocompatibility complex (MHC) a group of proteins that are combined with fragments of viruses or bacteria or other foreign substances and moved to the surface of cells so that they can be discovered by immune system cells

membrane channel a pore in a membrane, usually built of several proteins, whose function is usually to control the passage of substances into and out of the cell

messenger RNA (mRNA) a molecule that is transcribed from DNA and used as the template to create a protein

microRNA a small RNA naturally produced by cells that does not encode a protein. Many microRNAs bind to mRNA molecules to block their translation into proteins.

microtubule a fiber built of tubulin proteins that plays a key role in the transport of molecules through the cell and is used to build the mitotic spindle during cell division. Microtubules are a major part of the cytoskeleton, the scaffold of proteins that gives a cell its shape and structure.

monoclonal antibody an antibody that has been mass produced through the reproduction of a single cell

noncoding RNA RNA sequences or molecules that do not contain information used to make a protein

nonsense-mediated mRNA decay (NMD) a mechanism that cells use to destroy defective mRNA molecules

nuclear pore a basket-shaped opening in the membrane around the cell nucleus that helps to screen molecules to be moved in or out

nucleotide a subunit of DNA and RNA that consists of a base linked to a phosphate group and sugar

oncogene a gene that causes cancer if it becomes defective

phosphorylation a chemical modification of a molecule involving the addition of phosphate groups. This process is frequently used to pass signals within cells.

poly A tail a huge tail attached to the end of an mRNA molecule, made of long repeats of the subunit adenine, often 100 to 300 letters long

post-translational modification changes made in the last stages of the production of a protein, such as cutting off pieces or adding sugar molecules, which usually affect the molecule's functions

primary structure the most basic, "string" form of a DNA, RNA, or protein; its sequence

prion a small, misfolded protein fragment that causes mad cow disease and similar diseases. When it enters the body, it causes native prion proteins to misfold and spread.

protein a molecule made of amino acids, produced by a cell based on information in its genes

quaternary structure the highest level of a molecule's structure, defining the way it binds to and interacts with other molecules

receptor a molecule in a cell or on its surface that binds to a specific partner molecule, usually changing the receptor's activity, for example, by allowing it to bind to new molecules and activate a gene

recombination a process by which the DNA strand is cut and then the ends are rejoined at a different place, changing the order of elements in a DNA sequence as it is copied or passed down to offspring

retrotransposon a DNA sequence that is written into an RNA molecule, which is then used to re-create the DNA sequence and insert it into a new place in the genome

Rhesus factor (Rh factor) a protein used to type blood. Some people's blood cells contain Rhesus protein on their surfaces; other people's cells do not. If blood is exchanged between people with different types, the result will be a dangerous immune reaction.

RNA (ribonucleic acid) a molecule made of nucleotides that is produced by transcribing the information in a DNA sequence

secondary structure small three-dimensional shapes that form in molecules (usually referring to proteins) because of chemical attraction between subunits that lie near each other, making them fold

signaling pathway a sequence of chemical reactions that starts with a particular molecule, activates other molecules in a specific sequence, and ends at the same destination molecule, often leading to changes in the activation patterns of genes

small interfering RNA (siRNA) a short artificial RNA molecule designed to dock onto an mRNA and cause it to be destroyed, blocking its translation into protein

stop codon a three-letter sequence in DNA and RNA that signals the end of the protein-encoding part of a gene

structural biology the study of the three-dimensional architecture of cellular molecules

synchrotron a huge, ring-shaped instrument in which electrons and other particles are beamed along a circular track. As magnets bend their path, they release high-energy X-rays that can be used to examine the atomic structure of molecules or inorganic materials.

T cell receptor the structure on the surface of a T cell that recognizes fragments of foreign proteins, such as a molecule from a bacterium or virus bound to MHC molecules

tertiary structure the three-dimensional structure of a complete protein, made up of smaller secondary structures folded into functional modules, or domains

transcription the process by which DNA is read by RNA polymerase II or another molecular machine and an RNA strand is built based on its sequence

transfer RNA (tRNA) an RNA molecule that picks up a specific amino acid and delivers it to the ribosome so that it can be used in the assembly of a protein

translation the process by which a ribosome reads the information in an mRNA molecule and builds a protein based on its sequence

transposon a DNA sequence that can copy itself and move from one place in the genome to another, either by being physically cut out and pasted back in somewhere else or by first copying itself as an RNA that is then reverse transcribed into DNA

tubulin a protein subunit used to make microtubules

variable region a part of an antibody that has a unique structure in each cell because it is created by mixing and matching random parts of genes

vesicle a small membrane-enclosed compartment used to transport molecules through cells

Further Resources

Books and Articles

Atkins, Peter. *Atkins' Molecules.* 2d ed. Cambridge: Cambridge University Press, 2003. A beautiful book with lovely illustrations of molecules ranging from simple inorganic substances to complex organic molecules, such as hemoglobin. This is a very good introduction to chemistry that will be accessible to most high school students and adults.

Ball, Philip. *Stories of the Invisible: A Guided Tour of Molecules.* Oxford: Oxford University Press, 2001. Concentrating mostly on biological molecules, Ball gives a very readable, literary tour of the fundamental units of life and their functions in cells and organisms.

Beadle, George W. "Genetics and Metabolism in Neurospora." *Physiological Reviews* 25 (1945): 660. Written shortly after the discovery that "one gene encodes one enzyme," Beadle speculates on what his discoveries in mold have to say about the nature of genes just a few years before James Watson and Francis Crick discovered the structure of DNA.

Branden, Carl, and John Tooze. *Introduction to Protein Structure,* 2d ed. New York: Garland Publishing, 1999. A detailed overview of the chemistry and physics of proteins, for university students with some background in both fields.

Brown, Andrew. *In the Beginning Was the Worm.* London: Pocket Books, 2004. The story of an unlikely model organism in biology—the worm *C. elegans*—and the scientists who have used it to understand some of the most fascinating issues in modern biology.

Browne, Janet. *Charles Darwin: The Power of Place.* New York: Knopf, 2002. The second volume of the "definitive" biography of Charles Darwin.

———. *Charles Darwin: Voyaging.* Princeton, N.J.: Princeton University Press, 1995. The first volume of the "definitive" biography of Charles Darwin.

Caporale, Lynn Helena. *Darwin in the Genome: Molecular Strategies in Biological Evolution.* New York: McGraw-Hill, 2003. A new look at variation and natural selection based on discoveries from the genomes of humans and other species, written by a noted biochemist.

Carlson, Elof Axel. *Mendel's Legacy: The Origin of Classical Genetics.* Cold Spring Harbor, N.Y.: Cold Spring Harbor Laboratory Press, 2004. An excellent, easy-to-read history of genetics, from Gregor Mendel's work to the 1950s. Carlson explains the relationship between cell biology and genetics especially well.

———. *The Unfit: A History of a Bad Idea.* Cold Spring Harbor, N.Y.: Cold Spring Harbor Laboratory Press, 2001. An in-depth account of eugenics movements across the world.

Caudron, Maïwen, et al. "Spatial Coordination of Spindle Assembly by Chromosome-Mediated Signaling Gradients." *Science* 5,739 (2005): 1,373–1,376. An important scientific article that reveals some of the factors that help microtubules self-organize into the mitotic spindle during cell division.

Cavalli-Sforza, L. Luca, Paolo Menozzi, and Alberto Piazza. *The History and Geography of Human Genes.* Princeton, N.J.: Princeton University Press, 1994. For more than three decades Cavalli-Sforza has been interested in using genes (as well as other fields such as linguistics) to study human diversity and solve interesting historical questions such as where modern humans evolved and how they spread across the globe. This huge book is a compilation of what he and many researchers have found.

Chambers, Donald A. *DNA: The Double Helix: Perspective and Prospective at Forty Years.* New York: New York Academy of Sciences, 1995. A collection of historical papers by major figures involved in the discovery of DNA, with reminiscences from some of the authors.

Crick, Francis. *What Mad Pursuit: A Personal View of Scientific Discovery.* New York: Basic Books, 1988. Crick's account of dead ends, setbacks, wild ideas, and finally glory on the road to the discovery of the structure of DNA, with speculations on the future of neurobiology and other fields.

Darwin, Charles. *The Descent of Man.* Amherst, N.Y.: Prometheus, 1998. In this book, originally published 12 years after *On the Origin of Species,* Darwin outlines his ideas on the place of human beings in evolutionary theory.

———. *On the Origin of Species.* Edison, N.J.: Castle Books, 2004. Darwin's first, enormous work on evolution (published in 1859), which examines a huge number of facts while building a case for heredity, variation, and natural selection as the forces that produce new species from existing ones.

———. *The Voyage of the* Beagle. London, U.K.: Penguin Books, 1989. Darwin's account of his five years as a young naturalist aboard the *Beagle* is a scientific adventure story. He had not yet discovered the principles of evolution but was aware of the need for a scientific theory of life. Readers watch over his shoulder as he tries to make sense of questions that puzzled scientists everywhere in the mid-19th century.

Elliott, William H., and Daphne C. Elliott. *Biochemistry and Molecular Biology.* New York: Oxford University Press, 1997. An excellent college-level overview of the biochemistry of the cell.

Fruton, Joseph. *Proteins, Enzymes, Genes: The Interplay of Chemistry and Biology.* New Haven, Conn.: Yale University Press, 1999. A very detailed historical account of the lives and work of the chemists, physicists, and biologists who worked out the major functions of the molecules of life.

Gavin, Anne-Claude, et al. "Functional Organization of the Yeast Proteome by Systematic Analysis of Protein Complexes." *Nature* 415 (2002): 141–147. The first thorough survey of all the molecular machines at work in a living cell.

Gilbert, Scott. *Developmental Biology.* Sunderland, Mass.: Sinauer Associates, 1997. An excellent college-level text on all aspects of developmental biology.

Goldsmith, Timothy H., and William F. Zimmermann. *Biology, Evolution, and Human Nature.* New York: Wiley, 2001. Life from the level of genes to human biology and behavior.

Goodsell, David S. *The Machinery of Life.* New York: Springer-Verlag, 1993. Goodsell has devoted his career to giving people a new view of the molecular world, through amazing illustrations that present the cell as a busy, cosmopolitan place teeming with interacting molecules.

Gregory, T. Ryan, ed. *The Evolution of the Genome.* Boston: Elsevier Academic Press, 2005. An advanced-level book presenting the major themes of evolution in the age of genomes, written by leading researchers for graduate students and scientists.

Hall, Michael N., and Patrick Linder, eds. *The Early Days of Yeast Genetics.* Cold Spring Harbor N.Y.: Cold Spring Harbor Laboratory Press, 1993. For more than 70 years studies of yeast have provided crucial insights into the genetics and biology of plant and animal cells. This book contains personal accounts of early discoveries by pioneers in the field.

Henig, Robin Marantz. *A Monk and Two Peas.* London: Weidenfeld & Nicolson, 2000. A popular, easy-to-read account of Gregor Mendel's work and its impact on later science.

Judson, Horace Freeland. *The Eighth Day of Creation: Makers of the Revolution in Biology.* New York: Simon & Schuster, 1979. A comprehensive history of the science and people behind the creation of molecular biology, from the early 20th century to the 1970s, based on hundreds of hours of interviews that Judson conducted with the researchers who created this field.

Keller, Evelyn Fox. *A Feeling for the Organism: The Life and Work of Barbara McClintock.* San Francisco: W. H. Freeman, 1983. An account of the life and work of the discoverer of jumping genes, written before she received the Nobel Prize for her work. The book reveals the problems encountered by a brilliant, radical thinker—as well as a woman working in science during the middle of the 20th century.

Kohler, Robert E. *Lords of the Fly:* Drosophila *Genetics and the Experimental Life.* Chicago: University of Chicago Press, 1994. The story of Thomas Hunt Morgan and his "disciples," whose discoveries regarding fruit fly genes dominated genetics in the first half of the 20th century.

Lutz, Peter L. *The Rise of Experimental Biology: An Illustrated History.* Totowa, N.J.: Human Press, 2002. A very readable, wonderfully illustrated book tracing the history of biology from ancient times to the modern era.

Maddox, Brenda. *Rosalind Franklin: The Dark Lady of DNA.* London: HarperCollins, 2002. An account of the life and work of the woman who played a key role in the discovery of DNA's structure but who had trouble fitting into the scientific culture of London in the 1950s.

Magner, Lois N. *A History of the Life Sciences.* New York: M. Dekker, 1979. An excellent, wide-ranging book on the development of ideas about life, from ancient times to the dawn of genetic engineering.

McElheny, Victor K. *Watson and DNA: Making a Scientific Revolution.* Cambridge, Mass.: Perseus, 2003. A retrospective on the work and life of this extraordinary scientific personality.

Ptashne, Mark, and Alexander Gann. *Genes and Signals.* Cold Spring Harbor, N.Y.: Cold Spring Harbor Laboratory Press, 2002. A readable and nicely illustrated book presenting a modern view of how genes in bacteria are regulated and what these findings mean for the study of other organisms.

Purves, William K., et al. *Life: The Science of Biology.* Kenndallville, Ind.: Sinauer Associates and W. H. Freeman, 2003. A comprehensive overview of themes from the life sciences.

Reeve, Eric, ed. *Encyclopedia of Genetics.* London: Fitzroy Dearborn Publishers, 2001. A very interesting collection of essays on the major themes in genetics by researchers who are world leaders in their fields.

Sacks, Oliver. *The Island of the Colour-Blind and Cycad Island.* London: Picador, 1996. Neurobiologist Sacks's personal account of his travels to the Pacific islands of Cycad and Pingelap. There, he encountered people with an unusual genetic condition that allows them only to see shades of gray. The book contains his reflections on the impact of this condition on island culture.

Salem, Lionel. *Marvels of the Molecule.* New York: VCH Publishers, 1987. This small book is dedicated "to chemists and non-chemists alike" and is an excellent introduction to how atoms form molecules. Most of the book is devoted to inorganic molecules.

Stent, Gunther. *Molecular Genetics: An Introductory Narrative.* San Francisco: W. H. Freeman, 1971. A classic book for college-level students about the development of genetics and molecular biology by a researcher and teacher who witnessed it firsthand.

Tanford, Charles, and Jacqueline Reynolds. *Nature's Robots: A History of Proteins.* New York: Oxford University Press, 2001. A history of biochemical and physical studies of proteins and their functions and the major researchers in the field.

Tudge, Colin. *In Mendel's Footnotes.* London: Vintage, 2002. An excellent review of ideas and discoveries in genetics from Gregor Mendel's day to the 21st century.

———. *The Variety of Life: A Survey and a Celebration of All the Creatures That Have Ever Lived.* New York: Oxford University Press, 2000. A beautifully illustrated "tree of life" classifying and describing the spectrum of life on Earth.

Watson, James D. *The Double Helix.* New York: Atheneum, 1968. Watson's personal account of the discovery of the structure of DNA.

Watson, James D., and Francis Crick. "A Structure for Deoxyribose Nucleic Acid." *Nature* 171 (1953): 737–738. The original article in which Watson and Crick described the structure of DNA and its implications for genetics and evolution.

Web Sites

There are tens of thousands of Web sites devoted to the topics of molecular biology, genetics, evolution, and the other themes of this book. The small selection below provides original articles, teaching materials, multimedia resources, and links to hundreds of other excellent sites. All sites listed were accessed August 1, 2008.

American Society of Naturalists. "Evolution, Science, and Society: Evolutionary Biology and the National Research Agenda." Available online. URL: http://www.rci.rutgers.edu/~ecolevol/fulldoc.pdf. A document from the American Society of Naturalists and several other organizations summarizing evolutionary theory and showing how it has contributed to other fields, including health, agriculture, and the environmental sciences.

California Institute of Technology. "The Caltech Institute Archives." Available online. URL: http://archives.caltech.edu/index.cfm. This site hosts materials tracing the history of one of America's most important scientific institutes since 1891. One highlight is a huge collection of oral histories with firsthand accounts of some of the leading figures who have been at Caltech, including George Beadle, Max Delbrück, and many others.

Center for Genetics and Society. "CGS: Detailed Survey Results." Available online. URL: http://www.geneticsandsociety.org/article.php?id=404. This article presents the results of numerous surveys conducted in the United States and elsewhere on topics related to human cloning and stem cell research.

Dolan DNA Learning Center, Cold Spring Harbor Laboratory. "DNA Interactive." Available online. URL: http://www.dnai.org. This is a growing collection of multimedia and archival materials, including several hours of filmed interviews with leading figures in molecular biology, a time line of discoveries, an archive on the American eugenics movement, and a wealth of teaching materials on the topics of this book.

European Bioinformatics Institute. "2can." Available online. URL: http://www.ebi.ac.uk/2can/home.html. An educational site from the EBI, one of the world's major Internet

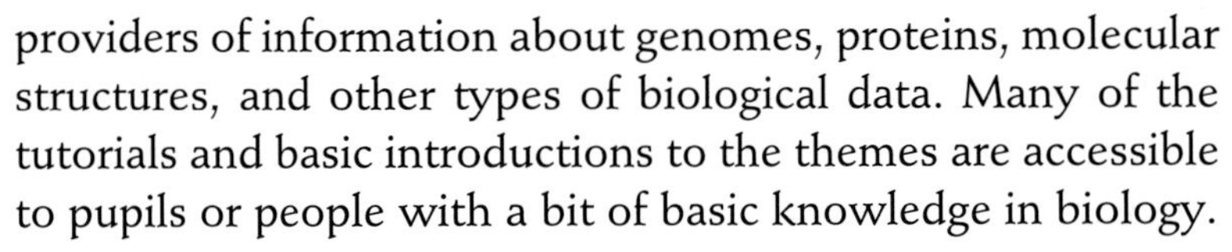

providers of information about genomes, proteins, molecular structures, and other types of biological data. Many of the tutorials and basic introductions to the themes are accessible to pupils or people with a bit of basic knowledge in biology.

Genetic Science Learning Center, University of Utah. "Learn Genetics." Available online. URL: http://learn.genetics.utah.edu/. An excellent Web site introducing the basics of genetic science, including the Biotechniques Virtual Laboratory; special features on the genetics and neurobiology of addiction, stem cells, and molecular genealogy; and podcasts on the genetics of perception and aging.

Maddison, D. R., and K. S. Schulz, eds. The Tree of Life Web Project. Available online. URL: http://tolweb.org. A site that has collected a huge number of articles and links from noted biologists on the question of assembling a "family tree" of life on Earth.

National Center for Biotechnology Information. "Bookshelf." Available online. URL: http://www.ncbi.nlm.nih.gov/sites/entrz?db=books. A collection of excellent books online ranging from biochemistry and molecular biology to health topics. Most of the works are quite technical, but many include very accessible introductions to the topics. Some highlights are *Molecular Biology of the Cell, Molecular Cell Biology,* and the *Wormbook.* There are also annual reports on health in the United States from the Centers for Disease Control and Prevention.

National Health Museum. "Access Excellence: Genetics Links." Available online. URL: http://www.accessexcellence.org/RC/genetics.php. Links and resources from the Access Excellence project of the National Health Museum.

Nobel Foundation. "Video Interviews with Nobel Laureates in Physiology or Medicine." Available online. URL: http://nobelprize.org/nobel_prizes/medicine/video_interviews.html. Video interviews with laureates from the past four decades, many of whom have been molecular biologists or researchers from related fields. Follow links to interviews with winners of other prizes, Nobel lectures, and other resources.

Research Collaboratory for Structural Bioinformatics. "RCSB Protein Data Bank." Available online. URL: http://www.rcsb.org. This site provides "a variety of tools and resources for studying the structures of biological macromolecules and their relationships to sequence, function, and disease." There is a multimedia tutorial on how to use the tools and databases. One special feature is the "Molecule of the Month," with beautiful illustrations by David Goodsell.

School of Crystallography of Birkbeck College, University of London. "The Principles of Protein Structure." Available online. URL: http://www.cryst.bbk.ac.uk/PPS. An online university program through which students can enroll to study structural biology on the Internet. The site hosts a wide range of basic introductory materials explaining the fundamentals of the field.

TalkOrigins. "The Talk Origins Archive." Available online. URL: http://www.talkdesign.org. A Web site devoted to "assessing the claims of the Intelligent Design movement from the perspective of mainstream science; addressing the wider political, cultural, philosophical, moral, religious, and educational issues that have inspired the ID movement; and providing an archive of materials that critically examine the scientific claims of the ID movement."

Tech Museum of Innovation, San Jose, California. "Understanding Genetics: Human Health and the Genome." Available online. URL: http://www.thetech.org/genetics. An excellent collection of news and feature stories on scientific discoveries and ethical issues surrounding genetics.

University of California, Santa Cruz. "UCSC Genome Bioinformatics." Available online. URL: http://genome.ucsc.edu/bestlinks.html. A portal to high-quality resources for the study of molecules and genomes, from UCSC and other sources. The map of the BRCA2 gene presented in chapter 2 was obtained from this site.

University of Cambridge. "The Complete Works of Charles Darwin Online." Available online. URL: http://darwin-online.org.uk. An online version of Darwin's complete publications, 20,000 private papers, and hundreds of supplementary works.

———. "Protein Crystallography Course." Available online. URL: http://www-structmed.cimr.cam.ac.uk/Course/. A basic online introduction to the use of X-ray crystallography in studying protein structures. The introductory articles are well written, with good metaphors and illustrations to explain this very complex topic.

Vega Science Trust. "Scientists at Vega." Available online. URL: http://www.vega.org.uk/video/internal/15. Filmed interviews with some of the great figures in 20th-century and current science, including Max Perutz, Kurt Wüthrich, Aaron Klug, Fred Sanger, John Sulston, Bert Sakmann, Christiane Nüsslein-Volhard, and others.

Index

Italic page numbers indicate illustrations.

G

H

I

N